Managing Computer Networks

A Case-Based Reasoning Approach

For a complete listing of the *Artech House Telecommunications Library,*
turn to the back of this book.

Managing Computer Networks

A Case-Based Reasoning Approach

Lundy Lewis

Artech House, Inc.
Boston • London

Library of Congress Cataloging-in-Publication Data
Lewis, Lundy.
   Managing computer networks : a case-based reasoning approach / Lundy Lewis
      p.   cm.
   Includes bibliographical references and index.
   ISBN 0-89006-799-6 (alk. paper)
   1. Computer networks—Management.   2. Case-based reasoning.   I. Title
ITK5105.5.L486   1995
004.6—dc20   95-33933
                                                                        CIP

British Library Cataloguing in Publication Data
Lewis, Lundy
   Managing Computer Networks:Cased-based
   Reasoning Approach
   I. Title
   004.6068

   ISBN 0-89006-799-6

© 1995 ARTECH HOUSE, INC.
685 Canton Street
Norwood, MA 02062

International Standard Book Number: 0-89006-799-6
Library of Congress Catalog Card Number: 95-33933

10   9   8   7   6   5   4   3   2   1

Contents

Preface

Unquestionably, computer networking will be recorded in history as one of the major scientific achievements of the twentieth century. In fact, it will likely be remembered as the most significant achievement of the last quarter of the century.

A computer network is broadly defined as a collection of devices and circuits that provides a means for transferring data from one computer to another. The computers on the network could reside locally in a city or a city block, or they could reside on opposite ends of the globe.

From a sociological perspective, computer networks allow people who are separated geographically and culturally to communicate and collaborate by means other than the traditional methods of post, telephony, facsimile, and published literature. This point has hit home with the author on numerous occasions over the last several years. For example, a co-author in Germany and I in the United States celebrated the acceptance of a paper in a journal with *wine bits* after a year of exchanging ideas, arguments, and algorithms over the Internet. We had the chance to meet face to face several months after the paper was published.

Despite their advantages, computer networks have spawned contentious ethical and legal dilemmas involving issues such as breach of security, slander, espionage, contractual binding, and responsible education of children. Indeed, there are legal cases concerning acts of pornography and seduction that have occurred over a computer network.

This book, however, focuses on a different sort of problem created by advances in networking during the last 25 years—the challenge of managing, maintaining, and servicing them. Those charged with this responsibility work behind the scenes to insure that networks are continuously operational and meet the expectations of the people who use them.

If I were to explain this book to a person on the street, I would say this: Whenever you get money from an ATM, there is a computer behind the screen that is talking to another computer (possibly thousands of miles away) making sure that you are who you say you are and that you have sufficient funds in your account. These computers reside on a computer network. One day, you will probably talk to a person who is a thousand miles away as though you were in the same room—via a computer network.

This book is about how to keep computer networks up and running—being able to anticipate potential problems and either prevent them or resolve them quickly with good solutions. More importantly, however, the book is about how to program a computer to perform these tasks automatically. Whether we say that these computers are smart or have *artificial intelligence* is not as important as rendering network maintenance easier and more efficient.

Consider that cars contain built-in mechanisms that assist us in maintaining and driving them. For example, most people have probably forgotten that at one time, cars' turning-signal levers did not return to place after a turn was made. Moreover, before that, there was a time when there were no turning signals at all. Only recently have we introduced mechanisms that learn how we drive and how we situate ourselves in our cars, thereby taking over some of these routine tasks for us. Indeed, at the University of Rochester in New York, students have built a programmable chip that learns one's style of switching gears on a multispeed bicycle and then automatically switches the gears for the biker. This grants the bicyclist time to concentrate on other details of biking.

This book is about building programs that learn how to maintain networks and solve network problems automatically, thereby granting network administrators and troubleshooters precious moments for other tasks.

Now, a more formal overview of the book is as follows:

Part I discusses (a) network management in general, with an emphasis on problem solving in network fault management, and (b) the expert system (ES) approach to automating human expertise. We argue that because networks are fluid (in that new devices, applications, users, and technologies are introduced as a matter of course), an ES in network management will require extensive maintenance and usually will become obsolete before its time. ESs are useful in stable domains where there are few surprises and where expertise in the domain is relatively fixed. Unfortunately, most networking domains are not like this.

Part II introduces a problem-solving paradigm that shows promise for alleviating the issues that surround ESs—case-based reasoning (CBR) systems. CBR systems are designed to learn and adapt in changing environments. The CBR paradigm of problem solving more closely approximates human learning and problem solving than the ES paradigm. CBR is described in detail and illustrated with examples in networking. In addition, case studies on successful CBR applications in domains other than networking are provided.

Finally, Part III brings the discussion down to earth, describing and instantiating an architecture for network fault management. CBR is just one component of the architecture, which also includes a network management platform (NMP) and a trouble ticket system (TTS). The fault management system is demonstrated using Cabletron Systems' Spectrum as its NMP and the Remedy Corporation's Action Request System as its TTS, with CBR embedded in it. In an alternative implementation, the CBR component is embedded into the NMP, and other network management applications may exploit its problem-solving expertise.

Acknowledgments

Thanks to Mom and Dad (Gladys and Lundy). Thanks to Mom particularly for the numerous discussions (nay, fights) on morals and religion. We hardly ever agreed on anything, but she taught me how to think and reflect. Thanks to Dad for teaching me about people, machines, and temperance.

Thanks to Dottie for her patience and impatience at the right moments during the preparation of the manuscript. She has a mind like a steel trap, and she was instrumental in getting me over some of the rough spots in the book. Her support was a great help.

My coworkers at Cabletron Systems and the general atmosphere at Cabletron have provided a good environment for trying to understand and solve outstanding problems in network management. The emphasis there is on innovative, useful networking solutions, as well as good business. I would especially like to acknowledge Bill Tracy, Utpal Datta, Russ Arrowsmith, Prasan Kaikini, and Carol Steele.

The reviewer of the manuscript unquestionably improved the book. I just hope I get the chance someday to critique her (or his) book.

PART I
Network Management and Automated Problem Solving

In Part I:

❏ *Network Management and Problem Solving*
❏ *Problem Solving with Expert Systems*

Part I describes the problem-solving expertise that is involved in network management and demonstrates the difficulties of representing this expertise in automated reasoning systems.

In Chapter 1, we describe the contents and scope of network management, placing particular emphasis on fault management and problem solving.

In Chapter 2, we describe a current, prominent vehicle for implementing problem-solving expertise—the ES. The ES framework is a generic method for representing problem-solving expertise in computer programs. There are many ES applications in network management. Some of these applications are prototype systems, while some are deployed in the field. We will examine the structure of an ES and show how ESs are developed and used in network management.

In addition, we will discuss some important issues that surround the development and maintenance of ESs:

- Knowledge representation;
- Knowledge acquisition;
- Brittleness;
- Learning and adaptability.

These tasks are challenging, and they impose restrictions on the use of ESs in domains that are in the midst of evolution, such as today's networks. Consequently, it is becoming increasingly difficult to develop and maintain ESs in network management, and alternative frameworks for representing problem-solving expertise are needed.

Network Management and Problem Solving   **1**

In Chapter 1:

❐ *What is Network Management?*
❐ *Network Management and Problem Solving*
❐ *Benefits of Automated Problem Solving in*
   Network Management

Our goal in this chapter is to understand the basic concepts and scope of network management. We highlight the practical task of solving network problems and discuss the prospects for capturing network problem-solving expertise in automated reasoning systems.

1.1 WHAT IS NETWORK MANAGEMENT?

A *computer network* is broadly defined as a collection of devices and circuits that provides a means for transferring data from one computer to another. The computers on a network can be used as stand-alone systems, or they can be used to send information to other computers and request and receive information from other computers.

Computer networks can be classified with respect to their field of coverage. A local area network (LAN) is comprised of a collection of interconnected computers that reside in a single room or building or in a cluster of buildings. A metropolitan area network (MAN) is a collection of interconnected LANs that are distributed over a broader geographical area—a city, for example. A wide area network (WAN) is a collection of interconnected LANs or MANs, and an internet generally refers to inter-connected LANs, MANs, and WANs that span the globe. (Technically speaking, an internet is simply a *network of networks,* whether in a single building or around the world. The Internet (with a capital I) has acquired the global connotation.)

Network management is the practice of (a) monitoring and control-ling an existing network so that the network stays up and running and meets the expectations of network users, (b) planning for network exten-sions and modifications in order to meet increasing demands on network operations, and (c) gracefully incorporating new elements into a network without interfering with existing operations.

There are a number of efforts under way to standardize network management and to render the practice of network management a science. For example, the International Standards Organization (ISO) Network Management forum has proposed the following functional model of net-work management, in which the subareas of network management are fault, configuration, performance, accounting, and security management.

Fault management provides facilities for detecting, isolating, and correcting problems on a network, including facilities to:

- Recognize network problems.
- Report and log network problems.
- Correlate and evaluate problem reports.
- Discover the cause(s) of problems.
- Initiate corrective procedures.
- Verify correction.

Configuration management provides facilities for collecting data on the number and kinds of devices on the network and the setup of each device and for modifying the setup of devices, including facilities to:

- Discover a network.
- Maintain a network inventory.
- Detect changes in the network.
- Execute changes in the network.

Performance management provides facilities for defining network performance metrics (including bandwidth utilization, device utilization, and intersegment and intrasegment traffic volume), evaluating network health in terms of these metrics, and comparing various network configurations with respect to network health, including facilities to:

- Collect raw network data, such as packet rate, packet collisions, and packet throughput.
- Translate the data into performance concepts.
- Evaluate the network under alternative configuration scenarios.
- Initiate healthier network configurations (also a part of configuration management).

Accounting management provides facilities for determining and allocating costs and charges to network users, including facilities to:

- Determine costs for network services.
- Collect data regarding usage of network services.
- Compile records of usage.
- Bill network users.

Security management provides facilities for protecting proprietary information against unauthorized intrusions, detecting and reporting successful and failed attempts of intrusion, and identifying intruders, including facilities to:

- Set policies for network usage.
- Establish and maintain encryption keys and authorization codes.
- Maintain a log of network access.
- Prevent and report unauthorized access.
- Initiate investigatory procedures for unauthorized access.
- Detect and prevent computer viruses.

The ISO definition is a good organizational principle and first approximation for understanding the scope of network management. However, researchers and practitioners in the networking industry typically point out other important subareas of network management:

Network design is the task of analyzing the requirements and goals of a network operation and outlining a network configuration that will satisfy these goals.

Network planning is the task of establishing the future goals of a network operation and designing a migration path from a current network configuration to a configuration that can support the future goals.

Network implementation is the task of overseeing the execution of a network design or a network migration plan.

Management of customer expectations is the task of defining metrics for qualities of service, insuring that customers receive the quality of service they demand and making corrections or allowances when customers' expectations are not met. This area also entails customer training and responding to customer comments and complaints.

Operation and maintenance budgeting is the task of prioritizing the services needed to maintain or modify a network and allocating expenditures accordingly.

Clearly, the practice of network management is a multidimensional, knowledge-intensive task. We are concerned with ways to automate the knowledge and expertise that are required to manage a network. While we cannot expect to replace good managers with robots, we should be able to build support tools that render managers' jobs less error-prone and costly, thereby increasing the efficiency of network management.

1.2 NETWORK MANAGEMENT AND PROBLEM SOLVING

It is important to note that a single network problem typically spans several network management subareas. For example, a problem of traffic congestion involves at least fault and performance management and generally involves several other subareas.

In this section, we formally define network problem solving and discuss a typical example that involves several subareas of network management.

1.2.1 What Is Network Problem Solving?

Most networking professionals have general ideas about what constitutes a network problem. However, general ideas and the language that we use to express them are notoriously slippery. That is, we do not always use words in precisely the same manner. In this section, we want to be rigorous about the words we use to discuss *network problems* and *network problem solving.* Thus, some definitions:

Definitions:

1. A network problem is an inference from a set of network symptoms, where the symptoms are the premises of the inference and the problem is the conclusion of the inference.

To illustrate, one might infer the problem, "server1 is inoperable" from the premises "client1 cannot access server1," "client2 cannot access server1," and "client3 cannot access server1." Or, one might infer the conclusion, "the bandwidth on link1 is too small" from repeated complaints about slowness of messages sent between the LANs that are connected by link1.

2. A network symptom is either a network event or a trouble report.
   a. A network event is a datum that issues from a network component or a network management system.
   b. A trouble report is the perception of a network abnormality that issues from a human user or observer of the network.

For example, network troubleshooters generally consider both network events and user complaints when they try to infer a network problem. If the troubleshooter sees a computer console message such as "permission denied," the symptom is a network event. When the troubleshooter hears a user say, "I cannot run this application anymore," the symptom is a trouble report.

Note that network events, which include computer console messages and management system messages, may not be detected or reported at all. For example, a network link going down is a network event, whether or not a component or a network management system reports it.

Trouble reports, which regard such matters as users' concerns about network sluggishness, their inability to use familiar applications, or their loss of contact with peer users, may be issued verbally, or they may be written down in a special language such as that used in a network trouble ticket.

3. The resolution of a problem is either the identification of a fault or an explanation.
   a. A fault is the physical or algorithmic source of a network symptom. Faults can be corrected.
   b. An explanation is a reason why a symptom is not a fault.

Once troubleshooters infer a problem from a set of network symptoms, they proceed to analyze the problem, seeking a resolution. Resolution might involve correcting a fault in software or hardware. For instance, suppose the cause of the problem, "server1 is down," is that server1 has core dumped. The temporary resolution of this problem is to restart server1, but the long-term resolution is to repair the code that caused server1 to core dump in the first place. Other common examples of faults in networking include bent pins, incorrect file permissions, and the inadvertent unplugging of devices.

On the other hand, the resolution of a problem might be an explanation. In this sense, a problem is simply explained away. The classic example of an explanation is *function as designed* (FAD). Another good example of an explanation of a problem is *user error.*

4. Problem solving is the process of inferring network problems from network symptoms and inferring resolutions from network problems.

For example, troubleshooters who monitor a network are typically presented with a wide range of network symptoms, where symptoms include network events and reports made by network users. Many symptoms (especially events) are routine and thus ignored.

From other symptoms, however, troubleshooters may infer a network problem. The problem, in turn, may trigger further reasoning, analyses, and tests in order to find a possible resolution. Sometimes the resolution is that the problem is really not a problem—for example, a FAD or user-error problem. Other times, the problem is indicative of a fault that can be corrected, thereby resolving the problem and eliminating the symptoms.

Many episodes of network problem solving are routine, involving such tasks as determining why a user is denied access to a remote system or why a user is unable to start a familiar application. Other episodes of problem solving are quite subtle, difficult, and expensive—for example, determining which course to take in a situation in which a group of users reports that their workstations are prohibitively slow and refuses to work until the network is repaired.

To represent problem-solving expertise in an automated reasoning system, one needs to know the range of possible symptoms, the range of possible problems, the range of possible resolutions, and the structure of the reasoning that is required to map symptoms to problems to resolutions. Experienced troubleshooters are adept problem solvers. However, if one were to witness an episode of problem solving and ask the trouble-shooter to describe his or her reasoning processes, one would find that an answer is not easily forthcoming.

As an illustration, the following section describes a moderately challenging episode of problem solving and provides the initial reasoning of an experienced troubleshooter in contemplation of the problem.

1.2.2 An Example of Network Problem Solving

The example in this section illustrates a relatively challenging, realistic exercise in network problem solving. Although the example is primarily a fault management problem, it reflects several other management areas.

Consider the scenario illustrated in Figure 1.1. The figure shows two networks, *N*1 and *N*2, connected by a link, *L*. There is a server, *S*, and a router, *R*, on *N*1.

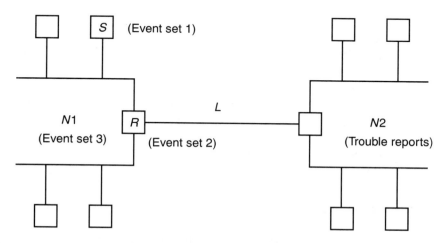

Figure 1.1 A sample problem-solving scenario in networking.

The network symptoms include the following events and trouble reports:

- *Event Set 1:* Server S logs a series of events indicating connection failures.
- *Event Set 2:* Router R logs a series of events indicating resource overload.
- *Event Set 3:* A probe on network $N1$ logs a series of events indicating traffic overload.
- *Trouble Reports:* Users on $N2$ log a series of reports indicating prohibitively sluggish behavior with client applications that access S.

The following text represents the initial reasoning of an experienced network troubleshooter who confronts the problem:

The problem is traffic congestion. Probably the best thing to do at first is nothing—wait a while to see if the problem goes away or becomes less severe. If this does not happen, there are several things that one might consider to correct the problem. For example:

1. Schedule the tasks of users on $N1$ and $N2$ to alleviate congestion.
2. Move the users on $N2$, who access S, to $N1$.
3. Move S to $N2$.
4. Duplicate S on $N2$.
5. Increase the bandwidth of $N1$.
6. Increase the resources of R.
7. Add an additional network $N3$ and move S and the users currently on $N1$ and $N2$ who access S to $N3$.
8. Redesign the client applications so that more data resides on the client, reducing the need for data traffic.
9. Some combination of #1–#8.

However, the best solution depends on several considerations:

If the maintenance budget is tight, then #1, #2, or #3 would be the best bet, since these options are inexpensive. But you have to consider whether #1 is realistic in the long run. With #2, you might resolve the troubles of the users on $N2$, but you have to consider the effect of this move on the bandwidth of $N1$. And if you choose #3, you will have to consider whether the users remaining on $N1$ who have to access S might begin to experience prohibitively sluggish behavior, in which case the same problem would arise all over again. Is #4 a practical possibility? Do you have license to install a duplicate server? Would the servers have to maintain synchronization? What would be the traffic overhead of this approach?

The options #5–#7 are expensive. If we choose #5, then we have not solved the problem of resource overload on R; and if we choose #6, we have not solved the problem of traffic overload on $N1$. If we choose #7, we have to worry about the effects on network performance when clients on $N3$ have to access other servers and applications on $N1$ and $N2$.

Is option #8 a possibility? Probably not in the short term...

This bit of reasoning demonstrates the first steps of mapping network symptoms to a problem and mapping the problem to alternative resolutions. We wish to capture the essence of such reasoning, generalize the reasoning so that it is applicable to similar problems, and cast the reasoning in a form that can be implemented in a computer program. If we are successful, we will have an automated problem solver that approximates the reasoning processes of the troubleshooter.

1.3 BENEFITS OF AUTOMATED PROBLEM SOLVING IN NETWORK MANAGEMENT

The benefits of codifying problem-solving expertise in a computer program are obvious. For example, the disadvantage of relying solely on the problem-solving abilities of an expert troubleshooter is simply that a network expert can have good days and bad days. Prior experiences of troubleshooting can be forgotten or clouded by current demands. Furthermore, the repair of a faulty network can be frustrating when an expert is off duty, on vacation, or absent for other reasons.

There are several tricks of the trade that troubleshooters use to support their problem-solving skills and make their jobs generally less difficult.

- *Maintaining and referring to records of experiences.* Most experts keep records of prior episodes of problem solving. Such records can take several forms. In the worse case, they are scribblings on a notepad. In the best case, they are organized and indexed in a file cabinet or a database system. This method helps to alleviate the problem of forgotten experiences but does not help in a situation in which the expert is absent.
- *Consulting experiences of other experts.* A complementary method is to consult with other experts. The obvious disadvantage of this method is that other experts may not be available.
- *Referring to procedures and algorithms in manuals and books.* Another complementary method is to study troubleshooting manuals to find procedures and tests that might be useful in devising a solution to a problem. This approach contributes to efficient problem solving, but it is not a substitute for experience.

These support systems are useful and indispensable. However, another method naturally suggests itself—to represent the problem-solving experience of a good expert in an automated reasoning system. The benefits of building such systems are as follows:

1. They do not forget, have bad days, or take vacations.
2. They can be used as a resource for less-experienced personnel.
3. They can decrease time and money spent on maintaining a network.
4. They can free up the experts' time so that they can work on other problems.

People who develop automated reasoning systems for network management typically use the expert system framework to represent problem-solving expertise. In Chapter 2, we will examine the ES framework and show how it is used in network management. In addition, we will look at some important issues surrounding the development and maintenance of ESs. In ensuing chapters, we will describe an alternative framework for problem solving.

1.4 SUMMARY

In this chapter, we described the general areas of network management and provided a formal definition of network problem solving. Problem solving is the process of mapping network symptoms to problems and mapping problems to resolutions. In addition, we argued that many episodes of network problem solving span several areas of network management. We illustrated this point with a problem that involves fault, configuration, and performance management; network planning; budget control; and the management of customer expectations. Moreover, we discussed the benefits of introducing automated problem-solving methods in network management.

1.5 FURTHER READING

The areas of network management are usually categorized into fault, configuration, performance, accounting, and security management. See any general text on networking or network management for discussions of these subjects. Good recent texts are Leinwand and Fang's *Network Management: A Practical Perspective*, Sloman's *Network and Distributed Systems Management*, and Terplan's *Communications Networks Management*.

Lewis's "AI and Intelligent Networks in the 1990s and Into the 21st Century" discusses the interrelations among the subareas of network management in the context of distributed artificial intelligence.

References that highlight the benefits of automated problem solving in network management and viable approaches to automated problem solving include Ericson, Ericson, and Minoli's *Expert Systems Applications to Integrated Network Management*; Goyal's "Artificial Intelligence in Support of Distributed Network Management" in *Network and Distributed Systems Management*; Liebowitz's *Expert System Applications to Telecommunications*; Aidarous and Plevyak's *Telecommunications Network Management Into the 21st Century*; Liebowitz and Prerau's *Worldwide Intelligent Systems*; Muralidhar's "Knowledge-Based Network Management" in *Telecommunications and Network Management Into the 21st Century*; and Goyal's "Knowledge Technologies for Evolving Networks" in *Integrated Network Management, II*.

Select Bibliography

Aidarous, S., and T. Plevyak, *Telecommunications Network Management Into the 21st Century,* New York: IEEE Press, 1994.

Ericson, E., L. Ericson, and D. Minoli (eds.), *Expert Systems Applications to Integrated Network Management,* Norwood, MA: Artech House, 1989.

Goyal, S., "Knowledge Technologies for Evolving Networks," in *Integrated Network Management, II,* I. Krishnan and W. Zimmer (eds.), Amsterdam: North-Holland/Elsevier Science Publishers, 1991.

Goyal, S., "Artificial Intelligence in Support of Distributed Network Management," in *Network and Distributed Systems Management,* M. Sloman (ed.), Wokingham, England: Addison-Wesley Publishing Company, 1995.

Leinwand, A., and K. Fang, *Network Management: A Practical Perspective,* Reading, MA: Addison-Wesley Publishing Company, 1993.

Lewis, L., "AI and Intelligent Networks in the 1990s and Into the 21st Century," in *Worldwide Intelligent Systems: Approaches to Telecommunications and Network Management,* J. Liebowitz and D. Prerau (eds.), Amsterdam: IOS Press, 1995.

Liebowitz, J. (ed.), *Expert System Applications to Telecommunications,* New York: John Wiley and Sons, 1988.

Liebowitz, J., and D. Prerau (eds.), *Worldwide Intelligent Systems,* Amsterdam: IOS Press, 1994.

Muralidhar, K., "Knowledge-Based Network Management," in *Telecommunications and Network Management Into the 21st Century,* S. Aidarous and T. Plevyak (eds.), New York: IEEE Press, 1994.

Sloman, M. (ed.), *Network and Distributed Systems Management,* Wokingham, England: Addison-Wesley Publishing Company, 1994.

Terplan, K., *Communications Networks Management,* 2nd Edition, Englewood Cliffs, NJ: Prentice-Hall, 1992.

Problem Solving with Expert Systems

2

Our goal in this chapter is to understand (a) the basic concepts of the ES framework of problem solving, (b) the strengths and weaknesses of the ES framework, and (c) applications of ESs in network management.

Before we begin our discussion, let us note that in the artificial intelligence literature there exists nomenclature that is closely related to the concept of an ES, including *production system, rule-based reasoning system, smart system,* and *knowledge-based system.* There are not clear-cut definitions or distinctions among these systems. They are best considered as a family of names that refer to approximately the same thing.

We should also note that the ES framework for problem solving is as much a paradigm for exploring human cognition as it is a paradigm for building practical problem-solving applications in industry. Although the question of human cognition is an intriguing topic, our discussion in this chapter focuses on practical applications of ESs in network management.

2.1 WHAT IS AN EXPERT SYSTEM?

In this section, we broadly define ESs and discuss how they work from an intuitive, commonsense perspective. In the following section, we will discuss the components of ESs and how they interact in more detail.

Definitions:

1. An expert system is a software system that mimics the problem-solving activities of a human agent with respect to a specific domain. Knowledge about problem solving in the domain is encoded as a set of *if-then* rules as opposed to the underlying causal structure of the domain.

   To illustrate, let us consider the domain of automobile mechanics and the problem, "car will not start." All of us have dealt with this problem at one time or another. A person who tries to solve the problem with if-then rules might reason as follows: If I turn the ignition and hear a click, then the battery is dead or the battery cables are corroded.
   Alternatively, a person who reasons about the problem with respect to the causal structure of the domain might think: When I turn the ignition, I am closing an electrical circuit that connects the following components—the ignition switch, generator, battery, starter, and regulator. Apparently, the circuit is disconnected somewhere. The question is where and why...
   Most of us approach problems initially with simple if-then rules like those described in the first example. If solutions are not forthcoming from our rules, we hand the problem over to someone who understands the causal structure of the domain or who at least has a better set of rules. Interestingly, these people generally try out their own set of rules before they reason more deeply about causal structure.
   The encoding of knowledge as if-then rules in an ES is a powerful method of representing several kinds of tasks in human problem solving.

2. Generic tasks that are candidates for ESs include:

   - *Monitoring:* observing, checking, and filtering observation data;
   - *Classification:* grouping observation data into higher level concepts;
   - *Prediction:* inferring possible future situations from current data and classifications;

- *Diagnosis:* investigating or inferring causes of conditions or problems;
- *Prescription:* identifying or inferring ways to correct problems;
- *Planning:* establishing a course of action to meet some objective;
- *Plan Repair:* rethinking and modifying a plan during plan execution;
- *Control:* directing, regulating, or coordinating systems or equipment.

As examples of some of these tasks, let us consider the domain of network fault management. A network troubleshooter generally sees a wide range of network events and trouble reports. Many of these symptoms, especially events, are ignored or cast aside as being of little consequence. At this stage, the troubleshooter is monitoring observation data—or simply observing, checking, and filtering the data.

The troubleshooter, however, may group some of the observation data into higher level concepts. Examples of such concepts are user-error, FAD, or "resource overload on router1 between 8:00 a.m. and 9:00 a.m." Note that at this stage the troubleshooter is inferring concepts from observation data, or, in the terms introduced in Chapter 1, inferring problems from symptoms.

At this juncture, the troubleshooter might proceed to predict the consequences of these concepts, posing questions such as: "Should I worry about the resource overload on router1 between 8:00 a.m. and 9:00 a.m.?" or "What overall effect might this have on the network enterprise?"

These examples illustrate the tasks of monitoring, classifying, and predicting. It would not be difficult to illustrate the remaining tasks of diagnosis, prescription, planning, plan repair, and control. For our purposes, however, it is enough to understand that these tasks can be simulated in ESs with the appropriate if-then rules.

3. The primary components of an ES are (a) a working memory (WM), (b) a rule base, and (c) an inference engine.

We will describe the components of ESs in detail in the following section. For now, let us try to get an intuitive grasp of these concepts.

A WM is a body of statements that one holds to be true in a domain. It is sometimes described as a *database of dynamically changing facts or data* or simply as a *domain database.* On an ordinary experiential level, my WM at a particular point in time might include statements such as the following:

- It's a sunny day.
- The time is 1:00 p.m.
- Aristotle is a philosopher.

Of course, my WM might be different ten minutes later.

A rule indicates statements that can be inferred from other statements. Examples of rules are, "If X is a philosopher then X is wise" and "If the time is 1:00 p.m. then it is time to get back to work." The *then* clause of a rule is a statement that could be included in the WM if the statement in the *if* clause were true.

Strictly speaking, however, note that a WM and a rule base by themselves do not actually give rise to additional statements. For example, suppose the WM contained the statement, "The time is 1:00 p.m.," and the rule base contained the single rule, "If the time is 1:00 p.m., then it is time to get back to work." What causes the system to infer the statement, "It is time to get back to work" and enter the statement into WM?

The inference engine in an ES is responsible for making such inferences. The engine looks at a rule base and the available statements in WM and executes the rules whose *if* conditions match the statements in WM (sometimes called rule firing).

In general, ESs solve problems by multiple iterations of the inference engine over WM and a rule base. On the first iteration, the inputs to the engine are a current WM and a rule base. The outputs of the engine are additional statements that feed back into the WM. On the next iteration, with a modified WM, the engine infers new statements that again feed back into WM and so forth. Figure 2.1 illustrates this basic cyclic operation of an inference engine.

For some tasks, the inference engine might have to execute only one cycle in order to find a solution to a problem or reach completion. In these cases, the ES reduces to a lookup table. A monitoring ES is a good example of this circumstance. The WM is a collection of observation statements in some domain. The rule base dictates which statements in the WM to discard and which to leave intact. The inference engine completes in one cycle over the WM and the rule base.

Other tasks might require multiple iterations of the inference engine. An example of this is an ES that incorporates monitoring, classification, and prediction in a single application.

At a gross level, the ES framework of problem solving appears to explain many of our ordinary problem-solving tasks. For example, our

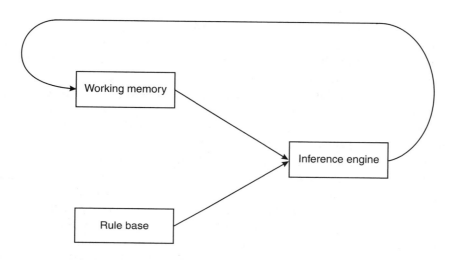

Figure 2.1 The cyclic operation of ESs.

WM at any point in time consists of what is delivered to us by our senses, plus other facts that we have accumulated and accepted in our years of experience. Our rule base consists of directives that we have learned to apply when we confront certain facts in WM. Our inference engine combines these facts and directives, resulting in actions that cause new facts to be asserted in the WM. We get along in the world by endless cycles of our inference engine over our WM and rule base. Our WM changes constantly, and our rule base evolves in tandem with our experiences.

When we begin implementing these concepts in ES applications, however, we will find that it can be difficult work. On one hand, if the domain of the application is well-defined and the rule base is correct, then ESs are quite useful and cost effective. There is little further work to do after development and testing in the field. On the other hand, if the domain is constantly changing—requiring rethinking of the rule base—then the development and maintenance of ESs can be frustrating and expensive.

2.2 DEVELOPING AND MAINTAINING EXPERT SYSTEMS

In the previous section, we examined the three main parts of an ES—the WM, the rule base, and the inference engine—and how they interact.

In this section, we will more closely investigate the major tasks involved in developing and maintaining ESs: scoping the domain, identifying mechanisms that will populate the WM, developing the rule base, and selecting an inference engine.

Many ES development packages, such as ES shells, provide tools that aid the developer in achieving these tasks. For example, some ES development packages provide tools for acquiring knowledge from domain experts. Other packages come off the shelf with a predefined set of inference engines; the developer can choose an appropriate engine for a particular application from this set. Other packages provide graphical interfaces for both the developer of the application and the user of the application.

It is also possible to build small ESs from scratch in a low-level programming language. Many useful ESs have been built in Fortran, C, or C++ and embedded as components in an existing operational system.

No matter whether the vehicle selected to develop an ES is a mature ES development package or a low-level programming language, one has to consider the tasks of scoping the domain, the means for populating the WM, developing the rule base, and selecting an inference engine. Therefore, this section will focus on these tasks.

2.2.1 Scoping the Domain

Precisely defining the domain for an ES application is an important first step in ES development.

In one sense, a *domain* is simply the area in which the ES is intended to be used. However, many ES projects have failed because the scope of the domain was too broad. For example, a project that proposed to develop an ES for network management in general is likely to fail, while a project that proposed to develop an ES for the management of faults in Cisco routers has a better chance of success.

Once the domain is sufficiently narrowed, the next task is to identify the sorts of problems and solutions with which experts in the domain must contend. This task commences the *knowledge acquisition* phase of ES development and is typically performed by extensive dialogue between a domain expert and a knowledge engineer.

Domain experts are well-versed in the sorts of problems they have to deal with and how to solve them but know relatively little about the problem-solving methods in ESs. Conversely, knowledge engineers are well-versed in ES problem-solving methods but typically know little about the domain under consideration.

The task of the knowledge engineer is to transfer the problem-solving skills of the expert into the ES framework. Some important questions that the knowledge engineer should consider to facilitate this transformation are these:

- What classes of problems is the ES expected to solve?
- What is the terminology used by the expert in characterizing these problems?
- Are there subproblems that have to be solved?
- What inputs does the domain expert use in order to recognize problems?
- What inputs does the domain expert use in order to solve problems?
- From what source does the expert derive these inputs?
- What are the classes of solutions to the problems?
- What is the terminology used by the expert in characterizing these solutions?

Note that the knowledge engineer can also appeal to other sources—such as textbooks and case studies of prior episodes of problem solving—to find answers to these questions.

Eventually, the knowledge engineer should be able to construct a kind of pseudo-domain language that represents the expert's problem-solving skills. The language can be considered as a cross between the everyday language used by the expert and the special syntax used in the ES. Some of the terms of the language will be used as inputs for the WM, and some of them will be used in the if-then rules that comprise the rule base. However, it usually turns out that the first approximation of the domain language differs considerably from the language of the final working system.

2.2.2 Populating the Working Memory

The WM represents an active database of current statements that hold true in the domain. The statements may be recordings of actual occurrences or events, actual data values, or statements that are generally accepted as being true. In some circles the WM is referred to as a *fact database* or *domain database*.

It is important to note that the WM is a dynamic structure, which means that its state changes over time. A WM may change as a result of the insertion of new statements, the deletion of existing statements, or changes to certain parameters' values.

We can group the sources of change in WM into two broad categories: external sources of change and internal sources of change. External sources of change reside outside the ES application. An example is a mechanism that monitors a domain and inserts statements in the WM in real time. Another example is a human user who answers questions posed by the ES, whereupon the ES modifies the WM according to the answers.

The internal source of change in the WM is caused by the ES's inference engine. Recall that the inputs to an inference engine are the current statements in the WM and the rule base. The outputs of the inference engine are insertions of new statements, deletions of statements, or modifications of parameters in the WM.

We will examine how an inference engine populates the WM below. Here, let us note that at some point the ES developer will have to decide which, if any, external sources will populate the WM. For example, many useful ESs do not have external sources (except the user interface) feeding the WM. These are sometimes called *offline* systems. The WM in other ESs are fed by real-time monitoring systems.

2.2.3 Developing a Rule Base

A rule is a representation of knowledge as a conditional, if-then clause. The antecedent of the rule (the *if* part) is a statement that can be true or false, and the consequent of the rule (the *then* part) is either (a) a command to assert a statement in WM, (b) a command to delete a statement from WM, (c) a command to update a parameter in WM, or (d) an action.

Actions include operational statements such as sending a message to a screen, asking the user a question, or rebooting a workstation. In addition, actions may be invocations of other ESs or calls to companion programs, such as a request for information from an external database.

An important task in developing an ES application is to build an initial rule base, recognizing that the rule base will likely be modified during the development, integration, and testing phases.

At this juncture, it will be useful to tell a story that has issued from the robotics community regarding the development of a rule base for a robot.

In the early 1980s, it was believed that one could implement the reasoning processes of a robot in a conventional ES. The inputs to the robot's WM would be the signals that were issued from the robot's sensors. However, it became clear that signals were not the right level of abstraction by which to reason about high-level terms such as chairs, tables, and walls.

The solution was to develop one small rule base that would map signals into signs (strings of 1 and 0s, for example) and another rule base that would map signs into symbols such as *chair straight ahead* and *table to the right*. Additional rule bases would then be required to represent knowledge about what to do in a situation in terms of these symbols, to map symbols back into signs, and to map signs into low-level signals that would drive the robot's effectors (see Figure 2.2).

Unfortunately, when all the rule bases were in place and a trial run was conducted, researchers found that the *amount of time* a robot required to reason what to do in a particular situation was quite disproportionate to *timely* behavior. By the time the robot figured out what it should do (i.e., by the time the robot mapped sensor data around the loop to control actions) the world had already changed, and the control actions were obsolete. Many jokes have been told about a robot having to reason for hours before making a small move that sometimes resulted in collisions with other objects in the vicinity of the robot.

The reaction to this problem was to recognize that a robot can make some inferences about control actions at the signal level without ever going up to the sign level and, similarly, can make some inferences about control actions at the sign level without ever going up to the symbol level. For example, see Figure 2.2. (In the figure, a rule base and inference engine is indicated by a rectangle.) At a low level of control, there exists a rule base that maps signals to actions, where the actions are also expressed in terms of signals. The sorts of tasks at this level require immediate reaction such as darting away from an approaching object. At an intermediate level of control based on signs, there exists a rule base that maps signs into actions, where the actions are also expressed in terms of signs; and at the top level of control, a rule base maps symbols into actions expressed in terms of symbols. For example, a task at the top level of control might be to plan a path from a starting point to a destination.

The verdict is still out regarding a *correct* architecture for autonomous control of a robot, and there is much debate on this topic in the robotics community. Nonetheless, we can glean from this story a useful way to categorize rules for use in ESs.

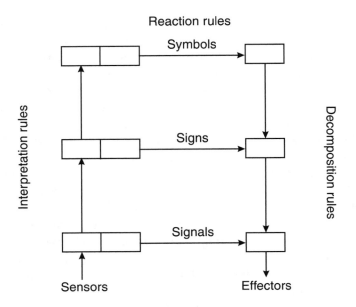

Figure 2.2 Multiple rule bases for robotic control.

Consider Figure 2.2 again. We may categorize rules into (a) interpretation rules, (b) reaction rules, or (c) decomposition rules.

Interpretation rules cause upward movement along the left side of the figure. These rules generally map low-level statements into higher level statements. An example in the network management domain is this:

1. If "the load of port A is greater that 40%" is true, then assert "the load of port A is high."

Reaction rules cause left-to-right movement. These rules generally map statements into actions, where both statements and actions are expressed at the same level of abstraction. An example:

2. If "the load of port A is high" is true, then "find load of port B."

Finally, decomposition rules cause downward movement along the right side of the figure. These rules generally map higher level actions into low-level actions. For example:

3. If "find load of port B" is true, then "get_value(B)."

Let us note that very few ESs require all of interpretation, reaction, and decomposition rules. For example, if the terms in the domain are expressed at just one level of abstraction, then interpretation rules are unnecessary. On the other hand, if the terms in the domain are expressed over several layers of abstraction, but the task of the system is classification only, then reaction rules and decomposition rules are unnecessary.

2.2.4 Selecting an Inference Engine

Generally, inference engines come in two varieties: forward-chaining engines (or event-driven systems) and backward-chaining engines (or goal-driven systems). To see the difference between these two kinds of engines, consider the following example:

A is true.
B is true.
If A and B are true then C is true.

A forward-chaining engine looks for matches among the antecedents of a rule and the facts in the WM. If a match is found, the engine executes the command in the consequent of the rule. In the example above, the statement "C is true" would be inserted as a new fact in WM.

A backward-chaining engine tries to determine whether the consequent or goal of a rule is true by ascertaining whether the antecedent of the rule is true. For example, if we were to pose the question, "Is C true?" the engine would see that C is true if A and B are true. Of course, the answer is yes.

It is easy to understand the operation of an inference engine with this simple example. In the presence of multiple rules in the rule base and hundreds of facts in WM, however, it is difficult to work through the operations of an inference engine. For example, in the case of a forward-chaining engine, what does one do if several rules are applicable?

To answer this question, let us subdivide the operation of a forward-chaining inference engine into the following subtasks: rule collection, conflict resolution, and rule execution. (Note: These subtasks are sometimes referred to as a search strategy.)

1. The rule-collection task collects all rules whose antecedents match statements in the WM. A *conflict* occurs when rule collection issues more than one rule.

2. The conflict-resolution task selects one rule from the collection to actually execute. Possible strategies of conflict resolution include:
   a. Select the first rule.
   b. Select the last rule.
   c. Select a rule randomly.
   d. Select the rule whose antecedent contains more information than any other rule.
   e. Measure the outcome of each rule according to some utility function and select the rule whose outcome affords the highest utility.
   f. Some combination of the above.
3. The rule-execution task executes the command in the consequent of the selected rule.

The subtasks of the inference engine are rather straightforward with the exception of the conflict-resolution task. Clearly, the strategy that one chooses will affect the operation of the ES. For example, if the strategy were to select the first rule in a collection of applicable rules, then the order of the rules in the rule base makes a difference.

We have described how inference engines work and pointed out some of the options that one may have to consider when designing an inference engine for an ES application. However, it is rare to actually have to build an inference engine from scratch, unless one is building a special-purpose ES in a programming language such as Fortran, C, or C++. Most ES development packages contain predefined inference engines that one can use in an application.

2.3 A SIMPLE EXPERT SYSTEM

It is instructive to work through the development of a simple ES in order to understand how ESs work. A note of warning is in order, however. The problem that our ES solves is essentially a *shortest path* problem (i.e., the problem of finding the shortest path between a starting point and a goal point). As such, the problem can be solved by methods other than ESs—for example, by using a greedy algorithm. Nonetheless, the exercise will illustrate the main components of ESs and how they interact.

(Note: Strictly speaking, the greedy algorithm does not actually *solve* the shortest path problem, because it frequently finds a suboptimal solution. In the artificial intelligence literature, such a solution is sometimes called a *satisficing* solution. Djikstra's algorithm and a number of others do solve the shortest path problem.)

2.3.1 Developing the System

Consider the 5 × 8 board shown in Figure 2.3. There is a single token on the board located in the start position (4,8). The problem is to find a path from the start position to the goal position (3,1) in the least number of moves. The token can move one square at a time: to the left, to the right, up, down, and diagonally. However, the token cannot move into closed squares and cannot move off the board.

Next, we show a representation of this problem in the ES framework. The domain language consists of a position $P = (X,Y)$, where (X,Y) are the Cartesian coordinates of P, and a set of all open positions in the set labeled *Open*. If a position is not a member of *Open*, then it is either closed or off the board.

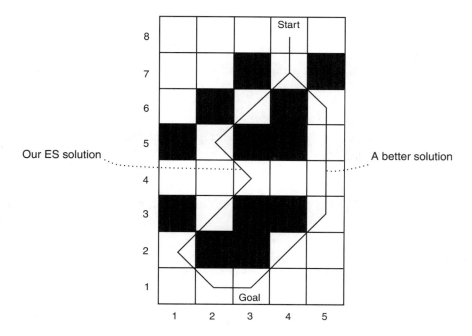

Figure 2.3 A problem for a simple expert system.

The Domain Language

$P = (X,Y)$
$Open = \{P1, P2,...\}$

Initial State of Working Memory

$P = (4,8)$
$Open = \{(1,8), (2,8), (3,8), (4,8), (5,8), (1,7), (2,7), (4,7), (1,6), (3,6), (5,6), (2,5), (5,5), (1,4), (2,4), (3,4), (4,4), (5,4), (5,3), (1,2), (2,2), (4,2), (5,2), (1,1), (1,2), (1,3), (1,4), (1,5)\}$

Goal State of Working Memory

$P = (3,1)$
$Open = \{(1,8), (2,8), (3,8), (4,8), (5,8), (1,7), (2,7), (4,7), (1,6), (3,6), (5,6), (2,5), (5,5), (1,4), (2,4), (3,4), (4,4), (5,4), (5,3), (1,2), (2,2), (4,2), (5,2), (1,1), (1,2), (1,3), (1,4), (1,5)\}$

The Rule Base

Rule 1. Moving left:
If $P = (X,Y)$ and $(X-1,Y)$ is a member of *Open* then retract $P = (X,Y)$ and assert $P = (X-1,Y)$.

Rule 2. Moving right:
If $P = (X,Y)$ and $(X+1,Y)$ is a member of *Open* then retract $P = (X,Y)$ and assert $P = (X+1,Y)$.

Rule 3. Moving up:
If $P = (X,Y)$ and $(X,Y+1)$ is a member of *Open* then retract $P = (X,Y)$ and assert $P = (X,Y+1)$.

Rule 4. Moving down:
If $P = (X,Y)$ and $(X,Y-1)$ is a member of *Open* then retract $P = (X,Y)$ and assert $P = (X,Y-1)$.

Rule 5. Moving up diagonally to the right:
If $P = (X,Y)$ and $(X+1,Y+1)$ is a member of *Open* then retract $P = (X,Y)$ and assert $P = (X+1,Y+1)$.

Rule 6. Moving up diagonally to the left:
If $P = (X,Y)$ and $(X-1,Y+1)$ is a member of *Open* then retract $P = (X,Y)$ and assert $P = (X-1,Y+1)$.

Rule 7. Moving down diagonally to the right:
If $P = (X,Y)$ and $(X+1,Y-1)$ is a member of *Open* then retract $P = (X,Y)$ and assert $P = (X+1,Y-1)$.

Rule 8. Moving down diagonally to the left:
If $P = (X,Y)$ and $(X-1,Y-1)$ is a member of *Open* then retract $P = (X,Y)$ and assert $P = (X-1,Y-1)$.

The Inference Engine

Repeat
  Collect applicable rules.
  Resolve conflicts:
  • Measure the Euclidean distance of each possible move to the goal position.
  • Select the move whose distance from the goal is smallest.
  • If distance = 0 then stop.
  • If there is a tie, select the first rule in the collection.
  Execute the rule.

Now, if one were to execute the ES on this problem, the solution would be the following path (see Figure 2.3):

(4,8) to (4,7) to (3,6) to (2,5) to (3,4) to (2,3) to (1,2) to (2,1) to (3,1)

Note that our inference makes nine iterations over the rule base and WM in order to find a solution to the problem.

2.3.2 Correcting the System

By inspection, we can see that there is at least one shorter path than the path found above:

(4,8) to (4,7) to (5,6) to (5,5) to (5,4) to (5,3) to (4,2) to (3,1)

Where did the system go wrong? It appears that the system could have made a better move than the move from (4,7) to (3,6). Had it moved from (4,7) to (5,6) instead, it would have found a better solution.

This simple observation illustrates some of the issues involved in the phase of maintaining and correcting an initial ES. That is, we have an ES that can provide a solution to the problem at hand, but we have discovered a better solution. If we cannot live with the results of the existing ES, our options are to revisit the domain language, the WM, the rule base, or the inference engine.

Clearly, our solution reflects an oversimplified inference engine: Besides not being able to find the best solution, there is no method for backtracking if the system finds itself oscillating around two moves. Some of the possible approaches to correcting the ES so that it finds better solutions are these:

- Modify the inference engine so that it looks ahead two or three moves instead of only one move. In this case, the inference engine would collect all the possible positions in a designated number of moves ahead and make a decision about where to move based on the closeness of those positions to the goal.
- Modify the domain language with a statement about board density with respect to P. Revise the rule base and the conflict resolution strategy taking board density into account.
- Simultaneously look for paths from the start position to the goal position and from the goal position to the start position. Note that this approach combines forward chaining with backward chaining. The inference engine would have to be modified accordingly.
- Implement Djikstra's algorithm in the inference engine, a method that guarantees to find the shortest path.

It is important to note that if we were to modify the ES so that it would find the best solution for this instance of the problem, we would have to worry about the sorts of difficulties we might run into when we execute the system on other instances of the problem. For example, the system might find itself following a dead-in path for some problems. Also note that the problem would be more difficult if we allowed the closed squares or the dimensions of the board to change erratically.

We will not work through these difficulties here, as the main point of this exercise is to get a basic understanding of the operation of ESs and the kinds of issues that we will have to deal with in developing and maintaining them. It is worth noting that very competent ESs have been developed for more complicated problems such as checkers and chess.

2.4 PROSPECTS FOR EXPERT SYSTEMS IN NETWORK MANAGEMENT

Since the 1960s, the ES framework of problem solving has been applied to numerous domains, including abstract domains for research purposes and practical domains that have yielded high payoff. As a result of this experience, the following issues are recognized as relatively hard problems involved in ES development and maintenance. These issues should be viewed as guidelines for the development of an ES application in network management.

2.4.1 Knowledge Representation

The ES framework is a hypothesis about knowledge and how humans solve problems. However, it is generally accepted that there are modes of knowledge and problem solving that do not fit well in the ES framework, including low-level reasoning over nonsymbolic representations such as sensory perceptions (e.g., recognizing faces) and learning and adapting as one engages in problem-solving activity. Alternative knowledge representation techniques that address these topics include neural networks (for representing sensory perceptions) and CBR systems (for representing learning and adapting during problem solving).

The problem of knowledge representation is that of finding a knowledge representation paradigm that fits the kind of task that is to be automated. Fortunately, most treatments of ESs provide a set of criteria for determining whether a candidate task is appropriate for an ES. However, there are no general methodologies for matching real-world tasks and the various knowledge representation schemes that are available to us. Currently, performing this task is as much an art as it is a science.

We should note that rather than reflecting an inherent problem with ESs, the problem of knowledge representation reflects the possibility of a mismatch between a real-world task and a representational scheme. For example, humans do not express sensory perceptions with if-then rules. If one were to force sensory perceptions into the ES framework without considering alternative representation schemes, one would run the risk of limiting the efficacy and success of the application.

2.4.2 Knowledge Acquisition

Recall that in ES development, there is usually (a) a knowledge engineer who knows the ES framework very well, and (b) an expert who knows a domain very well. The knowledge acquisition task is that of extracting the reasoning processes of the expert and representing the reasoning in the ES framework.

Knowledge acquisition can pose problems for the following reasons:

1. Experts are often unable to articulate the reasoning they use.
2. Experts do not remember all the details of how they solve a particular problem until they have to solve it.
3. Experts are so in demand to use their expertise on current problems that they cannot spare the time needed to work with knowledge engineers.
4. Experts are themselves learning and, therefore, do not know the solutions to all possible problems.
5. Knowledge engineers and experts may not "speak the same language." Consequently, it may be difficult for knowledge engineers to understand and capture expertise.
6. Unanticipated situations arise that the experts did not address.
7. The process of correcting an initial ES (e.g., by revisiting the domain language, WM, rule base, or the inference engine) frequently introduces other problems that are unanticipated.

There are several ways to alleviate the knowledge acquisition problem. Often, knowledge is compiled in texts and troubleshooting manuals, and knowledge engineers can refer to these works as their main or complementary source of knowledge. This approach is sometimes called *electrifying manuals.*

In addition, there are tools for knowledge acquisition that experts can use to organize knowledge. Many of these tools, which aim to minimize the need for knowledge engineers, produce first approximations of the domain language and the rule base for an application.

2.4.3 Brittleness

An ES is *brittle* if it does not recover gracefully when it confronts problems that it was not designed to solve. Generally, the application simply fails under such circumstances.

The causes of brittleness are related to the problems of knowledge acquisition: The system encounters a situation for which it does not have the relevant knowledge encoded in its rule base. Typical reasons for this include:

- The expert and knowledge engineer did not think of the situation when developing the rule base.
- The expert cannot describe how she solves the situation in a rule base structure.
- The expert does not know (ahead of time) how to solve the situation.
- The situation requires intuition (or luck), which is practically impossible to represent in the ES framework.

While these problems can occur in any domain, they are particularly troublesome in a domain involving a changing technology. In such domains, the expertise is not static. Experts are using their prior experience and learning new expertise as they go along.

One way around these sorts of problems is to try to develop new rules that cover the unforeseen situations in an evolving domain. Sometimes, however, such new rules interact with the old rules in such a way that the existing *correct* behavior of the system itself fails. This can also be a cause of brittleness.

2.4.4 Learning and Adaptability

Clearly, humans increase their problem-solving abilities with experience. In addition, good problem solvers are able to adapt their past experiences to new problems. However, it is difficult to design learning and adaptability into ES applications.

Despite the challenges involved in developing and maintaining ESs, those who are careful to bracket a task in network management that is well-understood and relatively constant have a good opportunity to build an ES that can automate the task and provide a useful service. Developers of successful ES applications in network management generally acknowledge that their success stems primarily from a domain that presents very few surprises and that the expertise required to solve problems in the domain is for the most part rigid and fixed.

Unfortunately, very few tasks in the management of today's networks are like this. Today's networks are fluid: New users and applications are introduced as a matter of course, network components are routinely updated with new functionalities, and new networking technologies such as switching networks are added to existing networks. This evolution in networking imposes a heavy burden on the people who have to maintain networks and makes it increasingly difficult to build expert support systems to help network maintenance personnel perform their jobs.

One may simply bear this burden or else seek other problem-solving techniques that show promise for learning and adapting in changing environments, much in the same way that good managers learn and adapt to new networking environments. The remainder of the book is devoted to a problem-solving method that shows promise for doing just this.

2.5 SUMMARY

In this chapter, we examined the ES framework for problem solving. We provided a formal characterization of ESs and discussed the steps involved in developing an ES application: defining a domain language, identifying sources that will populate the working memory, defining the rule base, and selecting the inference engine. As an illustration, we constructed a simple ES. Moreover, we discussed some issues in the development and maintenance of ESs that arise in domains that are inconstant and where problem-solving expertise is less than rigid: the problems of knowledge representation, knowledge acquisition, brittleness, and learning and adaptation. Because of these issues, we argued that ESs are inappropriate for many of today's network management tasks and that alternative problem-solving methods are needed. In the remainder of the book, we introduce CBR, an alternative method that shows promise to overcome these limitations.

2.6 FURTHER READING

A standard college text on artificial intelligence, including ESs, rule-based reasoning systems, and other methods of problem solving is Winston's

Artificial Intelligence. Books that focus on the development and maintenance of ES applications in industry are Buchanan and Shortliffe's *Rule-Based Expert Systems*; Hayes-Roth, Waterman, and Lenat's *Building Expert Systems*; and Waterman's *A Guide to Expert Systems.*

For practical applications of ESs and other AI methods see the book series *Innovative Applications of Artificial Intelligence,* initiated in 1989 by the American Association of Artificial Intelligence.

A good introductory paper on the use of ESs in network management is Cronk, Callahan, and Berstein's "Rule-Based Expert Systems for Network Management and Operations: An Introduction." Books that record various experiences with ESs in network management are Ericson, Ericson, and Minoli's *Expert Systems Applications to Integrated Network Management*; Liebowitz's *Expert System Application to Telecommunications; and Liebowitz and Prerau's Worldwide Intelligent Systems.*

Vesonder, Stolfo, Zielinski, Miller, and Copp's paper, "ACE: An Expert System for Telephone Cable Maintenance," demonstrates the beginnings of perhaps the most successful ES application in network management and describes some of the limitations of ES applications. Wright, Zielinski, and Horton's later paper, "The ACE System," sums up the experiences with ACE and provides guidelines for developing successful ES applications in network management.

Select Biography

Buchanan, B., and E. Shortliffe (eds.), *Rule-Based Expert Systems,* Reading, MA: Addison-Wesley Publishing Company, 1984.

Cronk, R., P. Callahan, and L. Berstein, "Rule-Based Expert Systems for Network Management and Operations: An Introduction," *IEEE Network Magazine,* Vol. 5, No. 4, September 1988.

Ericson, E., L. Ericson, and D. Minoli (eds.), *Expert Systems Applications to Integrated Network Management,* Norwood, MA: Artech House, 1989.

Hayes-Roth, F., D. Waterman, and D. Lenat (eds.), *Building Expert Systems,* Reading, MA: Addison-Wesley Publishing Company, 1983.

Liebowitz, J. (ed.), *Expert System Applications to Telecommunications,* New York: John Wiley and Sons, 1988.

Liebowitz, J., and D. Prerau (eds.), *Worldwide Intelligent Systems,* Amsterdam: IOS Press, 1994.

Vesonder, G., S. Stolfo, J. Zielinski, F. Miller, and D. Copp, "ACE: An Expert System for Telephone Cable Maintenance," *Proceedings of the Eight International Joint Conference on Artificial Intelligence,* Karlsruhe, West Germany, 1983.

Waterman, D., *A Guide to Expert Systems,* Reading, MA: Addison-Wesley Publishing Company, 1986.

Winston, P., *Artificial Intelligence,* Reading, MA: Addison-Wesley Publishing Company, 1992.

Wright, J., J. Zielinski, and E. Horton, "The ACE System," in *Expert System Applications to Telecommunications,* J. Liebowitz (ed.), New York: John Wiley and Sons, 1988.

Part II
Problem Solving with
Case-Based Reasoning

In Part II:

❏ *The Case-Based Reasoning Approach to Problem Solving*
❏ *Examples of Successful Case-Based Applications*

In Chapter 1, we described the general areas of network management and considered some of the typical problems that network managers face. We paid special attention to network fault management, and we argued that problem solving in fault management is ripe for the introduction of automated problem-solving techniques.

In Chapter 2, we described a useful framework for automating problem-solving expertise—ESs. However, we argued that the ES framework imposes limitations on the development and use of applications in network management. The ES framework is good for problem solving in domains that are constant and where the expertise for solving problems in the domain is relatively fixed.

Unfortunately, very few of today's network management tasks exhibit these characteristics. Today's networks are dynamic in the sense that new components, new networking technologies, new protocols, new applications, and new users are introduced routinely. Thus, an ES application that is suitable for a particular network operation will eventually lag behind the pace of network growth. The application will fail or at best offer the answer "I don't know" in response to new problems encountered in the management domain.

The limitations of the ES framework are generally grouped under the following headings: the knowledge representation problem, the knowledge acquisition problem, the brittleness problem, and the problem of learning and adapting in evolving domains. We discussed these issues in Chapter 2.

In Part II, we introduce a style of problem solving that has shown some relief for these limitations—CBR. Our goal is to demonstrate how to develop a CBR-based application that can be used in network fault management. We show how the application evolves elegantly and naturally as the network domain evolves and how the application demonstrates learning and adaptability as it faces and resolves new network problems.

In Chapter 3, we examine the CBR approach to problem solving. The chapter describes the architecture of a CBR system, the components of a CBR system, and various design options for implementing each of the components. In addition, the chapter provides examples of the design options that relate to network management.

In Chapter 4, we investigate four successful applications of CBR in domains outside of network management: banking, computer repair, manufacturing, and customer service. For each application, we will discuss the problem to be solved, prior approaches to solving the problem, the CBR approach, the benefits of the CBR approach, and the implications of the CBR application for network management.

The Case-Based Reasoning Approach to Problem Solving

3

In Chapter 3:

❏ *The Case-Based Reasoning Paradigm*
❏ *What is a Case?*
❏ *Case Retrieval*
❏ *Case Adaptation*
❏ *Case Execution*
❏ *Organizing Cases in a Case Library*

The legal domain provides a rich source for discussions about the logic of argumentation and, in particular, lends itself to the idea of CBR. The general CBR method of problem solving is related to a common way in which attorneys study their case books and develop arguments on behalf of their clients. To get an understanding of the way in which CBR works, let us look at the sort of preparations made by attorneys and their style of arguing when they represent a client in front of a jury or a judge.

Suppose a consumer purchased a product whose subsequent use caused irreparable damage to the consumer. The consumer proceeds to make a claim against the manufacturer of the product, and both the consumer and the manufacturer hire attorneys. Suppose the total set of circumstances involving the claim is represented as the set S. S might include the manufacturer's advertisements, warnings, instructions for proper and improper use of the product, and the specific use and expectations of the product by the consumer.

Let us think of ourselves as omniscient observers. Thus, we know the *truth*—we know the whole of S, we know the legal code, and we know whether the claim is well-founded. The first task of the lawyers, who are not omniscient observers, is to ascertain as much of the set S as possible. Their next task might be to refer to the current legal code to decide whether the claim is legitimate, given their partial knowledge of S. The notion of *referring to the legal code* can be taken literally—to cycle through the complete set of laws on the books and decide which laws are relevant to their particular subset of S. In civil cases, for example, the attorneys would study legislation that has to do with consumer protection, truth in advertisement, and consumer and manufacturer responsibilities.

However, this is not the way it is done in practice. Generally, each attorney goes through a collection of case books. A case book contains a listing of prior civil suits replete with the type of claim, the known circumstances, and the judgment. The attorneys try to find a case or number of cases whose claim and surrounding circumstances best match their client's claim and surrounding circumstances. If a close match is found, the style of the attorney's argument is, "the current case is like the case of such-and-such vs. such-and-such in which the judgment was found in favor of the defendant (or plaintiff)—thus this precedent suggests that my client should be judged accordingly."

There are several ways in which the suit could unfold from here. For example, a jury or judge might decide that the attorney for the plaintiff possesses more knowledge of S than the attorney for the defendant and that the attorney's citation of the case of such-and-such vs. such-and-such brings more to bear on the current case. The jury would then be inclined to favor the judgment of the case cited by the plaintiff's attorney.

On the other hand, a jury or judge might decide that the current case is special—i.e., that there are no prior cases that apply to the current case. It is inevitable then that the task of examining the legal code has to be undertaken. Possibly a judgment can be found by a careful analysis of existing legislation, but sometimes it turns out that new legislation has to be enacted or new legal terms defined in order to cover the new case. The final judgment in the new case is recorded in the case books and is used as a reference by attorneys that deal with similar cases in the future.

This sketch of the legal procedure illuminates the differences between the ES style of problem solving and the CBR style of problem solving. An attorney who begins developing an argument on behalf of a client by cycling through all the rules of the prevailing legal code and retrieving those rules that are applicable to the client's case is taking the ES approach. The attorney who cycles through case books and retrieves similar cases that have a bearing on the client's case is taking the CBR approach. In the legal domain, it is clear that the CBR style of arguing is more prevalent, and because the legal profession is one of the oldest professions, there is good reason to believe that there is merit to the CBR style of argument.

The underlying assumption of this type of legal argument is that one may undertake an examination of the legal code only if one has to. Initially, one can search for prior cases that have to do with the same sort of claim and exhibit similar surrounding circumstances and, if necessary, bring up the portion of the legal code that was shown to be relevant in the prior case. The justification for this style of argument is that the hard work might have already been done in a prior case, and one has to show only that the current case is sufficiently similar to the prior case. In this way, the legal system saves time and money and, at the same time, is reasonably confident that justice is served. In addition, legal experience becomes more robust and becomes increasingly fine tuned as cases are recorded in case books.

Before we look more closely at the generic CBR framework and how to apply it to engineering applications, let us examine a research project on the relation between CBR and legal procedure that was carried out by William Bain at Yale University.

Bain interviewed several judges in Connecticut to try to uncover their reasoning processes as they presided over legal cases. In one interview, Bain asked a judge what he would do if he were sentencing a 16-year-old for child molestation. The judge said that he was not sure whether he would grant the youth juvenile status or adult status. However, he was certain that if juvenile status were granted, the sentence would be three years of probation with psychiatric treatment. If adult status were granted, the sentence would be five years of probation with psychiatric treatment.

In an interview with the same judge about six weeks later, the judge had taken a different attitude toward the same case. Where before he was uncertain whether he would grant the youth juvenile status or adult status, this time he was decisive that he would treat the 16-year-old as an adult and give him the five years probation. When reminded of the difference in opinion in the earlier interview, the judge reflected that he had presided over a particularly agonizing juvenile case in the meantime, and this experience had affected his judgment. Upon further discussion and analysis, Bain formulated a general principle of legal reasoning to the effect that the features that one considers important in a case when making a determination are influenced by the experiences one has had dealing with those same features in similar cases.

Bain's observation suggests that a judge does not appeal to rules to make a determination about sentencing. Rather, the determination of a sentence in a particular case is a product of the judge's past experiences with similar kinds of cases, and what the judge considers salient about a particular kind of case can change over time. For this reason, the notion of a collection of explicit sentencing rules, such as what might be represented in an ES, is inappropriate. (As an aside, however, note that in societies where there is a concept of mandatory sentencing, it would be an easy matter to build an ES that would generate sentences automatically.)

The next phase of Bain's research was to create a computer model of subjective assessment over time and to implement the model in a computer program. The program was named JUDGE. For purposes of demonstration, the domain for JUDGE was limited to physical offenses that included murder, assault, and manslaughter. The input for the program was (a) a collection of prior cases having to do with physical offenses, (b) cases that JUDGE had examined and provided sentences for before, and (c) a current case waiting for a sentence. The output of the program was a sentence for the current case and an explanation for the sentence based on previous cases.

The language that Bain developed to describe physical offenses included *degree of threat, degree of force comparative to the opponent's force, intended harm,* and *degree of the achieved result.* Given values for these concepts, he developed formulas for approximating the degree of *heinousness* that accompanied an offense. Note that heinousness is a derived concept—it is an abstraction over the basic language in a case. Finally, given rules for mapping degrees of heinousness to degrees of punishment, JUDGE would suggest a sentence for an input case based on the sentences in prior, similar cases.

Bain's work illustrates two important components of CBR: a method for determining similarity among cases and a method for adapting old cases to fit new cases. The similarity of two cases is measured by comparing their values with respect to the concept of heinousness. A case is adapted by examining the relation between the degree of heinousness and the degree of punishment in the old case and, using this relation, to infer the degree of punishment for the new case.

In recent years, for other CBR applications, additional techniques have been introduced to compute similarity among cases and to adapt old cases to fit current problems. This chapter collects CBR techniques that have shown to be useful in practical engineering applications. As such, this chapter is like a CBR toolbox with instructions about how to use the tools. The choice of which techniques to use for a particular application depends upon a careful analysis of the domain and the kinds of problems that need to be solved.

3.1 THE CBR PARADIGM

CBR is a problem-solving method that offers potential solutions to the limitations of ESs. Some scientists argue that CBR is a model of cognition—that, at a gross level, it is really the way the mind works. Although this is a rich and interesting hypothesis, we will be more concerned about what we can achieve with CBR in practical engineering domains, particularly in the management and maintenance of computer networks.

The goals of CBR systems are:

1. To learn from experience;
2. To offer solutions to novel problems based on past experience;
3. To avoid extensive maintenance.

These are certainly desirable goals for any problem-solving agent. However, they are difficult to achieve. For one thing, the concept of *experience* is vague, and the concept of *learning* is equally vague. The CBR approach to problem solving that we study in this chapter will help us to form a notion of these concepts and apply them in practical applications. However, the reader should be forewarned that we cannot expect to fully understand and explain experience and learning, as these concepts are philosophically and psychologically complex.

The general CBR architecture is shown in Figure 3.1. The steps of a CBR system are to retrieve, adapt, execute, and organize episodes of problem solving. Former episodes of problem solving are represented as cases in a case library. When confronted with a new problem, a CBR system retrieves a similar case that might suggest a ball-park solution to the new problem. This step is called the *retrieval step*. The system then tries to modify the old solution in the prior case in an attempt to construct a specific solution for the new problem. This is called this the *adaptation step*. Then, the system executes the proposed solution and judges its efficacy in an *execution step*. Whatever the outcome—whether the solution works or not—the experience with the solution is embedded and organized in the case library for future reference. This is the *organization step*. As a result of the organization step, the case library evolves and becomes increasingly fine-tuned as the system solves problems.

The CBR framework for problem solving is an approximation of the commonsense notions of appealing to experience, applying experience, and getting experience, where experience is represented in the case library. The important questions that we want to uncover in this chapter are these:

1. What is an experience? How can an experience be recorded in a case? What are the dimensions of a case?
2. What does it mean to appeal to experience? How is this notion related to the retrieval of similar cases from a case library? What are the different methods of computing similarity?

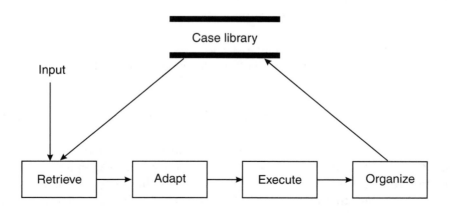

Figure 3.1 The general CBR architecture.

3. What does it mean to apply experience? How is this notion related to the adaptation of a case? What are the different kinds of adaptation?

4. What does it mean to get experience? How do people fit new experiences in with old experiences? How are these notions related to the organization of experiences with a case into a case library?

It is instructive to look at ordinary problem-solving activities as if they were CBR-like and determine what it buys us, or perhaps costs us. For example, let us imagine a database retrieval application for the personnel manager of a large company. The database contains profiles of each employee, including employment date, education, home address, skills, and salary. Suppose the personnel manager wants to know Jane Doe's salary. This is a common database query. However, we can look at the process as an example of CBR. The input case is:

employee = Jane Doe
salary = X

The database is the case library. We retrieve the case that matches the value of "employee" and map the value of "salary" in the retrieved case to the value of X in the input case.

It is doubtful in this example that the CBR perspective has bought us much. Suppose, however, that we are considering Jane for hire as a new employee and are wondering what her salary should be. This makes the problem more interesting and opens the way for a CBR approach to finding a solution.

To solve the problem, one considers Jane's profile in relation to the profiles of other employees in the company. Note, however, that one would not compare Jane and current employees with respect to their total profiles. Considerations such as gender, age, and marital status are irrelevant, whereas considerations such as years of experience, education, and skill set are relevant. If a close match along the latter dimensions were found, the salary of the current employee would be a reasonable ballpark figure to offer Jane. In addition, the ballpark salary might be tweaked to reflect the differences in her profile and the comparable profile. For example, if there were a very close match between the two except that Jane's set of skills was broader than the current employee, then one would be inclined to increase the offer.

At this juncture, let us begin our examination of CBR in detail. We wish to understand (a) the concept of a case; (b) alternative methods of case retrieval, case adaptation, and case execution; and (c) alternative principles for organizing cases in a case library.

3.2 WHAT IS A CASE?

As a first approximation, it is useful to liken a case to ordinary forms that we fill out in our everyday affairs—for example, forms that we fill out to take out loans, to apply to graduate schools, or to discover our psychological profiles. All forms contain a number of questions or slots, such as, "Are you buying or renting?" "What is your GPA?" or "What is your favorite color?" A completed form contains answers to these questions, or values for the slots.

To make the analogy between an ordinary form and a case explicit, let us provide the following definitions:

Definitions:

A form is a document with blank spaces for insertion of requested information. A completed form is a form for which requested information has been provided.

Analogously:

A case structure is a list of questions or slots. A case is an instance of a case structure, where answers are provided for the questions.

Thus, a blank form is analogous to a case structure, and a completed form is analogous to a case.

Let us note that forms do not explicitly ask what is really wanted. Often, we fill out forms so that somebody else can make a determination that is implicit in the form. When we fill out a form to take out a loan, we are never asked, "Are you a good risk?" or "Can you afford this or that?" These are questions whose values are derived from the values that we give to the simpler slots in the form, and usually there is a big, culminating question to which our answers build—for example, "Loan approved?"

3.2.1 Three Major Components of a Case

To understand the major components of a case, let us continue our analogy with everyday forms.

A form exhibits three primary components:

1. A primary slot that is the main question;
2. A number of basic slots that represent background information;
3. A method or formula that is used to determine an answer to the main question, given the background information.

If we are filling out a form, we do not see the answer to the main question until sometime later. Generally, we are only aware of the basic slots for which we provide answers. We do not see the method or formula for determining the answer unless we ask for an explanation.

On the other hand, if we are administering a form so that we ourselves can make a determination, we know mainly about the formula. Given our client's answers to the basic slots, we apply the formula in order arrive at an answer to the main question.

If we reflect a little on our dealings with forms and how we use them, we can get a good idea of the concepts of a case structure and a case in CBR. A case structure is like a blank form, and a case is a completed form. However, to fully appreciate the CBR framework for problem solving, we have to advance to thinking of a case as representing a particular experience—for example, an experience of solving a problem, achieving a task, classifying an item, or devising a plan.

The structure of an experience consists of the same components of a form. There is a predicament in which we find ourselves that calls for some answer or some action. The need for an answer or action is analogous to the main question. The predicament is analogous to the basic slots that will record background information. The way in which we determine how to get out of the predicament is analogous to the formula that we use to determine an answer to the question.

With effort, we can describe our everyday experiences in terms of these three general categories, although it is difficult. Specifically, people often find it quite difficult to articulate how they have gotten out of predicaments, or how they have discovered a solution to a problem.

Nonetheless, since a case is a vehicle for expressing an experience, it is important to make the categories explicit. For example, a form that exhibits an inadequate structure for use in a CBR system is shown in Figure 3.2. This figure shows a standard form for documenting network maintenance problems. The main question is represented by the slot that asks for the summarization of the maintenance problem. The background information is represented by the slots that ask for the name, address, and location of the user.

However, the answer to the question appears to be restricted to the replacement or updating of hardware or software. Clearly, there are other kinds of solutions for network maintenance problems. Also, it is important to note that in this form one would be hard-pressed to find something analogous to a formula or method by which to determine an answer to the main question. There is no slot that explicitly asks for the method that the troubleshooter used to resolve the maintenance problem.

Figure 3.3 shows a simple revision of the form that renders the important components of the problem-solving activity more explicit.

```
Failure Date _____   Report Number _____
Network User _____   Telephone _____
Network Address _____   Location _____
Briefly summarize the maintenance problem _____
_____
_____
_____

Hardware or software replaced _____
_____
Hardware or software updated _____
_____
Is the problem now completely resolved? _____
Is the network user satisfied? _____
Any other comments? _____
_____
_____
_____

Person completing this form _____
Date _____   Telephone _____
```

Figure 3.2 A sample network maintenance form.

Failure Date _____   Report Number _____
Network User _____   Telephone _____
Network Address _____   Location

Briefly summarize the maintenance problem _____

_____

_____

Summarize the solution to the problem _____

_____

_____

Is the problem now completely resolved? _____
Is the network user satisfied? _____
Describe the methods used (test, manuals, other trouble reports) _____

_____

_____

_____

Figure 3.3 A revised network maintenance form.

The main question is represented by the slot "Briefly summarize the maintenance problem" and the answer is represented by the slot "Summarize the solution to the problem." The slot "Describe the methods used" holds the reasoning that leads to the answer for the main question.

3.2.2 Developing a Case Language

The initial task in developing a CBR application is to construct a suitable language for talking about the domain of the application.

Definition:

A case language is the enumeration of the names of slots in a case structure and (optionally) an enumeration of the legitimate values or expressions that may be entered in the slots.

For example, the structure in Figure 3.3 qualifies as a case language, albeit a rather sparse one. Note that we placed no restrictions on the kinds of values that can fill the slots.

It is important that the language of a case at least allows one to specify the three major components of a case: the big question (typically posed as "The Problem" or some variation thereof), other slots that record background information, and a slot that holds the method used to arrive at a solution to the problem. We took a first step towards constructing such a language in Figure 3.3.

However, if the language that we construct for a case is underconstrained, we can run into difficulties. The more liberal we are about the language that can be used to describe experiences with network problems, the more likely we are to misinterpret the nature of the problem, the solution to the problem, and the method used to find a solution. The form in Figure 3.3 reflects such an underconstrained language, since we allowed the values of the slots to be expressed in freestyle text with no restrictions. This if fine if everybody who deals with the form speaks the same jargon, but this is hardly ever true.

A better approach is to impose restrictions on the case language so that one can be precise about the experience that is represented in the case. For example, we may choose to categorize network problems into general types of problems, where each type is further subdivided into subtypes. Under this approach, we place restrictions on the sorts of things that can be said to describe network problems.

The disadvantage of this approach is that our language can become overconstrained. If the language is overly precise and restrictive, then we might not have the means to express an important piece of information if is not included in the language. The form in Figure 3.2 is overconstrained to some extent, because it assumes that network problems are resolved by updating or replacing hardware or software.

There is no easy recipe for devising an ideal case language for a CBR application. In the beginning, the best approach we can take is to devise a preliminary language that is a happy medium between an underconstrained language and an overconstrained language. However, if we provide ways to allow the language to evolve as new experiences with the network occur, then the preliminary language will converge toward an ideal language.

3.2.3 Case Structures Center around Themes

Definition:

The theme of a case structure is an organizational principle that affords some degree of coherence among the case slots.

Generally, the theme of a case structure is indicated by the slot that represents the main question. For example, if we are designing a form for a loan application, we try to figure out the background information that is required to determine an answer to the main question "Loan approved?" Thus, the main question serves as the organizational principle that groups the remaining slots in the structure.

On the other hand, we can imagine questions other than "Loan approved?" that we could answer by looking at the slots and values in a completed loan application. Some of the questions that come to mind are "What is the socioeconomic status of this person?" and "What are the hobbies of this person?" We would look at the same form, but pick the slots that are relevant to the question of interest and use a different formula to derive an answer to the question.

Given a form with many slots, then, it is possible that some of the slots are organized around one theme and other slots are organized around another theme. This is especially true in psychological testing forms where different personality traits are determined by different subsets of slots in the form. The importance of this observation is that forms can lead multiple lives and, thus, be used for multiple purposes. This also holds true for the concept of a case in CBR.

3.2.4 Where to Find Case Libraries

Definitions:

A case library is an apparatus that holds a number of individual cases (e.g., a database or a collection of ASCII files).

A seed case library is a case library that holds a relatively small number of cases containing broadbrush sketches of problems and solutions that occur in a domain.

A likely place to find a case library is in a domain for which there exists a well-established form, where the form includes slots that can be interpreted as the main question, background information, and a method for answering the question. Such a form will have shown its value by predicting more often than not an answer to the question for which it was designed. Our jobs as engineers are easier but perhaps less interesting if there exists an explicit formula for deriving an answer to the question.

At the other extreme, it is possible, but more difficult, to design a new case structure for an application and then collect individual cases to populate a seed case library. The first task would be to develop a case language. We would have to ask, "What is my main question?" "What sorts of questions would I ask that would lead to an answer to the question?" and "What is the formula for deriving the answer?" These questions are challenging, and it is difficult to devise a correct language in the beginning of development. However, the CBR paradigm allows a seed case library to converge toward a better, richer library as the application is used.

There are other potential case libraries that we can exploit in engineering applications. If we have a question we want to answer, and there are no forms that lead to an answer to the question, then we can look for a form used in a different context that might contain enough information to answer the current question. This follows from the fact that we can think of a case structure as containing multiple themes. Of course, we would have to select the basic slots in the structure that can be used to answer the question and ignore the rest; and we might have to add a few more slots. In addition, we would have to come up with the formula for deriving the answer.

3.3 CASE RETRIEVAL

We can get an approximation of appealing to experience by thinking of it as a retrieval of a prior case from a case library. Simply put, we try to find a prior case whose problem description exhibits maximum similarity to the problem description in a current case. We examine the solution that was used in the old case and try to determine whether the same solution would work in our new situation. Alternatively, we may choose to look at the method that was used to arrive at the old solution and try the same method in our new situation.

Now, however, we have introduced a rather vague concept by talking about maximum similarity. If we can settle on a good notion for determining when two cases are similar, then we are halfway toward understanding what it means to appeal to experience. In this section, we will examine four techniques for computing similarity between cases:

- Key-term matching;
- Relevance matching;
- Structure matching;
- Geometric matching.

Because this chapter is a CBR toolbox, we can consider each technique as an option that we can use if our application calls for it.

3.3.1 Key-Term Matching

Definitions:

Key-term matching is a case retrieval technique in which matches are sought between an input case and cases in a case library with respect to a predefined list of key terms.

A key term is a linguistic expression that is commonly used and recognized in a given domain language.

An obvious, easy way to retrieve useful cases from a case library is to look through the library and select those cases that match on the problem description. The requirement for taking this approach is that we have a well-structured language for categorizing different kinds of problems. For example, the free-form style of describing problems in Figures 3.2 and 3.3 would not work unless we added another slot whose values are chosen from a predefined list of key terms for describing network problems. If we had such a list, then we could perform the matching routine over the key terms.

Alternatively, one could look for occurrences of key terms in the freestyle summarization of the problem. The following list represents a good start toward devising a list of key terms for describing network problems.

- Communications;
- Quality of service;
- Software;
- Hardware;
- Environment;
- Maintenance;
- Upgrade;
- Relocate;
- Printer;
- Workstation;
- PC;
- Down;
- Slow;
- Heavy;
- Locked;
- Inoperable;
- Boot;
- Access.

Note that this approach to case retrieval can be rather haphazard. For example "printer" and "down" matches with "The temperature in the printer room is down to 40 degrees."

Another difficulty that we will have to deal with is the possible fuzziness of the values for a case slot. What do we do when we are comparing the values "very heavy" and "moderately heavy"? Clearly, we do not want to say that there is no match at all between these values, but determining what to say is problematic.

There are some practical ways to get around the problem of fuzziness. One way is to disallow the possibility by being sufficiently restrictive when we construct the language for describing the domain. Another way is to accommodate the possibility of fuzzy values in our retrieval mechanism by brute force—for example, by returning some value over comparisons such as "heavy," "slightly more than heavy," "slightly less than heavy," and so forth. However, this approach can get messy if we are not careful. There is a body of work on the concept of fuzziness and fuzzy inference that we can appeal to in order to come up with a more elegant solution to the problem. (See the references at the end of the chapter.) These techniques show promise. However, at this stage, they are experimental.

3.3.2 Relevance Matching

Definitions:

Relevance matching is a case retrieval technique in which matches between an input case and cases in a case library are guided by a set of relevance rules.

A relevance rule is a structure of the form, "If A then B," where A represents a problem type, and B indicates a set of parameters that are required to find a solution to A.

We are commonly taught that one of the important tasks in problem solving is to separate out those facts that are pertinent to the problem from those that are not; and the ability to do this is perceived as genius. On an ordinary experiential level, we have learned to do this quite well. If we stop to think about the abundance of our sense impressions that present themselves to us at any moment of experience, we see that we put the majority of those impressions out of mind and focus on just those impressions that are relevant to the task at hand. When we drive our cars on a busy interstate highway, we do not focus on surrounding natural beauties that otherwise would hold our attention. We focus on road signs, what is in front of us, and what is coming up behind us or beside us.

What are the underlying cognitive mechanisms that allow us to learn to select a few important facts from an abundance of them? Answering this question is difficult, but it is important to attempt to do so in a discussion of CBR systems. Accordingly, in this section, we will try to get a handle on how we are able to focus on just the important facts when we undertake tasks. We can use this approximation for practical CBR applications.

For most of our CBR applications, there will be facts recorded in the case that will be irrelevant to the problem at hand. This can happen for several reasons. First, we have argued that it is desirable to interpret an ordinary form as a case structure. When a collection of completed forms is available, we are likely to have a case library at our fingertips that we can exploit by applying some of the CBR methods for case retrieval and case adaptation. However, if we are so lucky in finding domains for which there already exists a body of experience, it is inevitable that the case structure will contain extraneous information (since these forms were designed for other tasks). Some of the information will be important for achieving our task, while some of it will be insignificant.

Second, we have shown how case structures can contain multiple themes. Some of the information in the structure will be relevant to one theme, and other information will be relevant to another theme. With these sorts of applications, we are faced again with the problem of selecting information that is relevant to the task at hand and disregarding everything else.

Finally, it will often be true that our construction of a case language for talking about a domain is, to some extent, underconstrained. In most domains, there are several kinds of problems that we will want to solve. If we restrict ourselves to a language that circumscribes just one kind of problem, we are limited with respect to other problems. However, different problems that one may encounter in a domain may require different sets of information. Thus, we would want to design a case structure that holds all the information we would ever need, even though we would never need all the information for any one instance of problem solving. This, again, is an example of the problem of knowing how to select the right information when we need it.

For now, let us presume that we have an underconstrained language for describing a domain. Given two cases, then, it seems reasonable to think of similarity as a function of the number of matches of slot/value pairs over the two cases, where the larger number of matches indicates increasing similarity. However, it is easy to see that this method of computing similarity can get us into trouble. This is because we would be performing our matching routine over the relevant data as well as the irrelevant data. It would be quite possible, then, that two cases would be determined highly similar for the wrong reasons.

(As an aside, note that the phenomenon of "declaring similarity for the wrong reasons" explains why many jokes and cartoons are funny. A mouse and a kangaroo are similar in several respects. But see what happens when a cat tangles with a kangaroo.)

The task of focusing on just the right data in a problem is simple on the surface. If we know the slots in a case that are relevant to a particular kind of problem, then we record this knowledge in a relevance rule. A relevance rule tells us what kind of information would be sufficient to determine a solution to the problem. Note that a relevance rule does not dictate what the most similar case will be or what we should do with a similar case when we find it. In this way, a relevance rule differs from the notion of a rule in an ES. Rather than offering a solution to a problem, a relevance rule tells us where to start looking for it.

If we let the total number of slots in a case be represented by the set S, the general form of a set of relevance rules is the following:

If the problem is of type P_1 then match on subset S_1 of S.

If the problem is of type P_2 then match on subset S_2 of S.

...

If the problem is of type P_X then match on subset S_X of S.

The retrieval mechanism first appeals to the appropriate rule to find the subset of slots to be concerned with. Then, it proceeds to execute the matching routine over the case library where matching occurs over just the subset of slots.

A variation on this approach is to devise a function that maps the values of the subset of slots S_X to a single aggregate value in an abstract slot. It is an easy matter, then, to perform the matching routine over this abstract slot. Recall that this is the approach that was taken in the program JUDGE discussed earlier in the chapter, where heinousness was created as the abstract slot.

This method of case retrieval has good points and bad points: On one hand, it allows us to come to grips with a basic notion of similarity that we can use in applications. In addition, the technique allows us to exploit useful kinds of problem solving over a body of experience in a case library that initially might have been designed for other purposes. In other words, given an existing case library, we can get more for our money by complementing it with a set of relevance rules.

The outstanding question regarding the method is "Where do relevance rules come from?" (This question is often referred to in the CBR literature as the indexing problem.) This is a hard question that we will not try to answer completely in this section. The safest approach, and the approach that is used most often in current CBR applications, is to manually handcraft and test a set of relevance rules. In Chapter 4, we will examine a technique for deriving a set of relevance rules automatically from an existing case library and fine-tuning the rules as the case library evolves over time.

The manual construction of relevance rules has its advantages. In several applications, it has been shown that experts who are good at problem solving in their domain find it relatively easy to articulate a good set of relevance rules. In contrast, they find it much more frustrating to articulate a set of rules for general problem solving that can be used in a rule-based ES. The reason is clear. At this juncture of CBR application development, they do not have to articulate the exact ways in which they solve their problems. For most of us who have sought help from experts about a particular problem, we can probably identify with responses like "If I had your problem, I would first look at such-and-such and such-and-such." This sort of knowledge does not offer much help for developing an ES application, but it is just what we want in the initial stages of developing a CBR application.

3.3.3 Structure Matching

Definitions:

Structure matching is a case retrieval technique in which matches between an input case and cases in a case library are sought with respect to case structure.

A case structure is a representation of the relations of the slots in a case.

(Note: The concept of a case structure in this section should not be confused with the concept of case structure defined in Section 3.2.)

We saw in the previous section that a set of relevance rules can help us retrieve useful cases from a case library by matching on the slots of cases that are relevant to a particular problem. However, an additional burden stems from developing and maintaining the relevance rules. If we can do away with the task of formulating a set of relevance rules but still be able to focus on the right information for a particular problem, we will have achieved a great deal.

In this section, we will develop a method of case retrieval that does not appeal to relevance rules. Moreover, the method will suggest techniques for developing increasingly richer languages for describing an application domain. Unfortunately, there is a price that we have to pay for it.

Our goal is to provide a way to pick out the relevant features of a case by looking primarily at the structure of the case and secondarily at the content of the case. Many of us first began to understand the notion of structure when we learned how to diagram sentences in grammar school. Recall that we learned diagrammatic conventions by which we could represent the interrelations between parts of a sentence. For example, consider the following three sentences:

1. Tom hit the ball with the bat.
2. Tom hit the ball with the stripes.
3. Tom hit the ball with the stick.

On the surface, it is clear that the three sentences are equally similar. However, we know that the pair of sentences 1 and 3 are more similar than the pair 1 and 2 or the pair 2 and 3, because 1 and 3 share the same structure and the other pairs have a different structure. The prepositional phrase in 1 and 3 is adverbial, whereas in 2 it is adjectival. The structures of 1 and 3 and the structure of 2 are shown in Figure 3.4.

In the same way, the method of case retrieval based on structure matching captures the idea of determining similarity by looking at structural similarities of two cases rather than looking at their surface similarities.

We will first describe the machinery needed to understand the notion of structure matching, and then we will see how we can apply the method in practice. The formalism that we will use roughly follows methods in object-oriented programming and design, although other formalisms such as logic programming or semantic networks may be used instead.

First let us define the following terms: object, message, class, attribute, relation, and concept.

Definitions:

An object is any particular thing to which one can ascribe an identity, a set of attributes, and a set of behaviors.

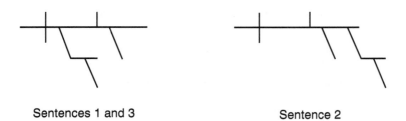

Sentences 1 and 3   Sentence 2

Figure 3.4 A comparison of sentence structures.

A message is query or statement that issues from one object to another object.

A class is a generalization over a set of objects, where each object is an instance of the class.

A particular workstation, for example, has an identity that is usually its name. The attributes of a workstation are its static characteristics—whether it is turned on or off, the applications installed on it, other devices to which it is connected, and its CPU load. The behavior of a workstation refers to its characteristics when it reacts to messages from other objects and when it sends messages to other objects—for example, the console messages that occur when one tries to run an application on the workstation or when one tries to log onto the workstation, and the messages that occur when the workstation tries to connect to a remote workstation.

A class holds those characteristics that would be exhibited by any object that is an instance of the class. A workstation class might have slots for an identity, its attributes, and its behaviors. However, the class itself would not have values for these slots unless default values were known to be true of any particular workstation.

Note that a case language for a CBR application could contain several classes of objects and several instances of these classes—for example, classes representing users, problems, solutions, workstations, networks, peripheral devices, applications, and communication protocols.

In addition to attributes, we may talk about an object in terms of its relations to other objects or how an object fits into high-level concepts that describe the domain:

Definitions:

An attribute is a one-place predicate that is ascribed to an object.

A relation an n-place predicate that is ascribed to n objects, where $n > 1$.

A concept is an n-place predicate that is ascribed to attributes or relations, where $n > 0$.

If we have an object in a case that is a physical thing, one of the first kinds of attributes that comes to mind is color. Other examples of attributes of physical things include weight, height, length, and volume. The attributes of a workstation include the kind of computer, the name of the computer, RAM, operating system version, and IP address.

Some of the kinds of relations that can occur between two objects are "on," "inside," "part-of," "has-part," "connected-to," "a-kind-of," "heavier than," and "more expensive than." The most common relation that can hold among three objects is "between." Typical examples of relations in the networking domain are the following:

- workstation1 is on subnet1;
- subnet1 contains workstation1;
- router1 is between subnet1 and subnet2.

Concepts are predicates that apply to attributes and relations. Examples of concepts include "It ought to be that" and "It is requested that." For example:

- It ought to be that workstation1 has operating system 4.1.5.
- It is requested that workstation1 has operating system 4.1.5.

Importantly, concepts can be used to represent the dynamics among a set of objects. A common example is the concept "causes":

"router1 is between server1 and client1" and "router1 throughput is very high" causes "response time of client1 is very slow."

In this example, the concept is the two-place predicate "causes." The first place of the predicate is the conjunction of a relation and an attribute, and the second place of the predicate is an attribute.

At this point in our discussion, we only note that these distinctions show the lengths to which we can go in developing a case language for an application. We may choose to represent the objects in the domain in terms of attributes only; we may choose to represent them in terms of attributes and relations; or we may choose to represent them in terms of attributes, relations, and concepts. However, when we introduce the notion of concepts, we impose increasingly more structure on our descriptions of the domain.

Given these distinctions, we lay down the principles for determining similarity based on structure matching as follows:

- Matches over attributes are not good grounds for declaring similarity.
- Matches over relations afford better grounds for declaring similarity.
- Matches over concepts afford the best grounds for declaring similarity.

Intuitively, these principles are palatable. Often, surface similarities between two objects deceive us into believing that there is a real similarity between the objects. This is commonplace in the consumer world when a quality object is copied with respect to surface similarities but not with respect to structural similarities. If we are informed enough to compare the objects with respect to structural similarities, we might be able to expose the differences. The same kind of phenomenon happens in problem-solving activities when on the surface, two problems might appear similar, but upon further examination, we find that they are not similar.

To apply these principles to the case retrieval task in CBR systems, we would first collect a set of cases from the library that matches an outstanding, unresolved case with respect to the problem type. Then, we would prune this set by picking out those cases whose concept descriptions match the concept description in the unresolved case. If no matches over concepts are forthcoming, we prune the set by picking out the cases whose relational descriptions match the relational description in the unresolved case. Finally, if all else fails, we pick out those cases whose attributes match the attributes of the unresolved case.

To illustrate, consider an unresolved case that records the problem "workstation1 cannot access workstation2" and consider the following possible pieces of information in the description of the case:

1. Workstation1 is on subnet1.
2. Workstation1 is on subnet1 and workstation2 is in on subnet2.
3. Workstation1's attempt to access workstation2 causes the message "Permission denied" on workstation1 and the message "Access attempted" on workstation2.
4. Workstation1 is an SGI workstation.
5. The mode of access between workstation1 and workstation2 is telnet.

We are more likely to find good, targeted solutions to the problem if there are cases in the library that exhibit the concept in #3. Short of this, it seems reasonable that solutions might be found in problems that exhibit #2, #5, or #1, in that order. It is less likely, albeit possible, that solutions might be found in cases that exhibit #4. As a last resort, we would be presented with all cases that matched with respect to the problem type.

The price that we have to pay to use this method is that it is hard to describe network problems with so much precision, hard to devise languages that reflect such precision, and hard to be consistent in using the language to describe the problems. Furthermore, there is extra burden on the retrieval algorithm. In some domains, however, it is possible to catalogue a set of concepts that describe interesting situations in the domain. When we can do this, the method can offer some ingenious solutions to difficult problems.

3.3.4 Geometric Matching

Definitions:

Geometric matching is a case retrieval technique in which matches between an input case and cases in a case library are determined by geometric distance.

Geometric distance is a numeric measure of similarity between two cases, where smaller distances indicate increasing similarity.

In our examples so far, we have considered slots whose values are symbolic or numeric. In this section, we describe a method of computing similarity when values are solely numeric. We will not find many cases like this in real applications, but in some domains it will be possible to translate symbolic values to numeric values and apply the method. When we can do this, we have the raw material for a useful method to compute a measure of similarity between an input case and each case in the case library.

The method involves using the general formula for calculating the geometric distance between any two points in a graph, where smaller distances indicate increasing similarity. Consider a simple example. Suppose P and Q are two cases, each with two slots A and B and their numeric values x and y, respectively:

$$P: \ A = x_P$$
$$B = x_P$$

$$Q: \quad A = x_Q$$
$$B = x_Q$$

We can measure the degree of similarity $d(P,Q)$ between P and Q with the following formula:

$$d(P,Q) = [(x_P - x_Q)^2 + (y_P - y_Q)^2]^{1/2}$$

To illustrate with a simple example, suppose the case describes a piece of notebook paper and the salient features of the paper are width and height, both of which are measured in inches. Then the two pieces of paper shown in Figure 3.5—an 8 × 11 sheet and an 8 × 14 sheet— have a degree of similarity of 3. The 8 × 11 and a 5 × 7 sheets of paper shown in Figure 3.6 have a degree of similarity of 5. Thus, all else being

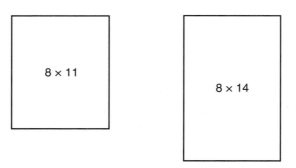

Figure 3.5 Two cases with a similarity measure of 3.

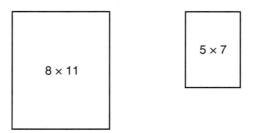

Figure 3.6 Two cases with a similarity measure of 5.

equal, we can say that the 8 × 11 sheet of paper is more similar to the 8 × 14 sheet than to the 5 × 7 sheet.

It is straightforward to generalize this notion for arbitrarily multiple slots in cases. If the values are numeric and measured in like units, we can think of a case as a point in n-dimensional space, where the dimensions represent the slots of the case and the case is the point whose coordinates are determined by the values of the slots. A case library, then, is a collection of points plotted in the n-dimensional space. If we were to look at a graph of the total number of cases in two- or three-dimensional spaces, we could identify clusters of cases that are similar and those cases that are more or less unique. Furthermore, given any input case, we can retrieve the case that is most similar to it by computing geometric distances.

There are several variations on the method of computing geometric distance that we might find useful for engineering applications. For example, a particular slot in a case—A, for example—might be more important than all other slots. Therefore, we would modify the formula so that we decrease the contribution of the value $(x_P - x_Q)^2$ when we compute the total similarity of two cases, since smaller distances indicate increasing similarity (often called weighted geometric distance). On the other hand an application might require that a value of a slot in two cases be identical or that a special relation holds between two or more slots. For these kinds of applications, we modify the basic formula for computing distance accordingly.

There are other issues that we will have to face when using geometric distance as a basis for determining similarity between two cases. In the discussion so far, we have assumed that the features of a case are measured in like units, for example in inches. It is a different matter when the units of multiple slots are incommensurable, for example, inches, pounds, volume, and age. If two cases vary over a few units with respect to volume but vary over thousands of units with respect to age, then the determination of similarity based on the basic formula of geometric distance is clearly biased with respect to volume. The usual way around this difficulty is to normalize each feature in a case library around some interval, say [0,1]. With these kinds of problems, we execute a normalization routine over the case library and our input case before we start the retrieval process.

Finally, performance issues might arise when we have thousands of cases in the case library. It would be expensive to find the most similar case by comparing the similarity measure between an input case and each case in the case library one by one. One way to cut down the computational expense is to perform a preprocessing step over the case library using a Veronoi-type algorithm.

The Veronoi algorithm divides the space of cases into hypercubes, where a hypercube is a subspace over the dimensions of the case. The space is divided recursively into smaller hypercubes until no hypercube contains more than a predefined number of cases, say 2. Figure 3.7 illustrates the method. Given an input case, we go directly to the hypercube that contains the case and compare it to the cases that are contained in the hypercube. For some retrieval tasks, we might have to look at the cases in bordering hypercubes.

In sum, the notion of similarity based on geometric distance provides us with a useful, flexible method of case retrieval when slot values are numeric, but it has limitations and difficulties that we will have deal with in most real-world applications. Some of the questions we should ask when considering this method for particular applications are:

1. Is it realistic to think of the slot values of a case to be exclusively numeric?
2. If not, is it realistic to translate the values into numeric values?
3. Can we modify the basic formula of computing geometric distance to reflect the relative importance of case slots?

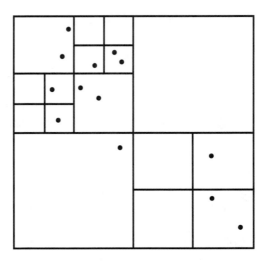

Figure 3.7 Division of a space of cases into hypercubes.

4. Will we have to normalize values of multiple slots over some common interval when the units in which the values are measured are incommensurable?

5. Can we rest with the performance of a one-by-one comparison between an input case and each case in a case library, or should we use other methods to speed up the performance?

3.4 CASE ADAPTATION

For some CBR applications we might want to stop with the retrieval step. The retrieved cases that record prior episodes of problem solving serve as simple reminders or suggestions as to how we might go about solving a current problem, and we as problem solvers do the rest. The old problems and their solutions set our minds in motion. After some thinking and meditating, we might convince ourselves that an old solution to a problem applies directly to the current problem, or we might determine that with some minor adjustments, we can make the old solution apply to the current problem.

In this section, we will examine some techniques whereby CBR systems make such minor adjustments automatically. The CBR term for making these adjustments is case adaptation. From a commonsense perspective, case adaptation is related to the notion of applying experience. The techniques we will describe are:

- Null adaptation;
- Adaptation by substitution;
- Parameterized adaptation;
- Critic-based adaptation.

3.4.1 Null Adaptation

Definition:

Null adaptation is a case adaptation technique in which a solution in a retrieved case is mapped directly to a new case, with no adjustments or modifications to the solution.

Null adaptation is a good way to start the learning process of CBR systems. In the beginning, when we put a CBR system in the field, we will have to plant a seed case library. The cases in the library will offer solutions to some general problems that occur in the application domain. It is likely that the seed library will be rather sparse and immature, and we will have no choice at first but to map an old solution to a new problem as a proposal. If the solution works, we can be satisfied. If a solution does not work or if it works with some minor adjustments, we should have the means to record this experience in the case. In this way, we can avoid making the same mistake in the future and, at the same time, increase the coverage of the library and make the system more useful for future problem solving.

A good way to record an experience with a proposed solution to a problem is to introduce additional slots in a case that make the experience with the proposed solution explicit. For example:

Solution:
  Proposed Solution:
  Solution Status:
  Reason if No Good:

When an input case is first described, these slots will be empty. When a prior case is retrieved, the solution in the prior case will occupy the slot "Proposed Solution" in the current case. Finally, the user records the experience with the proposed solution in the slots "Solution Status" and "Reason if No Good." If the "Solution Status" is good, then the proposed solution is mapped to the slot "Solution."

In the networking domain, we can imagine a problem, "application1 will not execute on workstation1," for which the proposed solution is "upgrade operating system on workstation1," but where it turns out that this solution is unacceptable because other applications running on workstation1 mandate the current operating system. This experience is recorded in the case library as follows:

Problem: application1 will not execute on workstation1

Solution:
  Proposed Solution: upgrade operating system

Solution Status: No Good
Reason if No Good: current operating system mandated

This case is entered in the library. When the same type of problem occurs the next time around, the system first looks at the cases in which the proposed solution is "upgrade operating system" and the solution is "no good" and tries to determine whether the current case is like that case. In our example, the system might pose the question "current operating system mandated?" before it proposes the solution. If the answer is yes, the solution in the case would not be recommended; otherwise, the case would generate the proposed solution "upgrade operating system."

3.4.2 Adaptation by Substitution

Definition:

Adaptation by substitution is a case adaptation technique in which one or more pieces of a solution in a retrieved case are replaced by new pieces.

A common way to adapt an old, unworkable solution to fit a new problem is to substitute pieces of the old solution with new pieces that make it workable. We can think of many times that we needed a tool to perform some task, and the correct tool was unavailable or the tool that we had was inappropriate. Our first thought is to try to find a substitute that will do the job. An example of this is when we try to extract a nail with a pair of pliers instead of a hammer, or when we serve decaffeinated coffee instead of caffeinated coffee. Similarly, in the programming world, we are taught that it is good practice to consider the available predefined functions that perform a computation before we rewrite the function from scratch.

The important question, however, is to know where to look for substitutes. A good way to think of this question is to imagine that a particular piece of a solution belongs to a more general category that contains the piece as an instance. We can presume, then, that the category will contain other instances that might do the job for us. Since the instances belong to the same category, it follows that they will be sufficiently alike to render it reasonable to consider them for the task. In the commonsense example above, the category might be "nail extracting tools," where instances include a hammer, a crowbar, pliers, hands, and so forth.

A useful variation on this method is to impose a structure on the instances of a category so that we are more likely to find a good substitute quickly. A simple kind of structure is ordering. If the instances of a category were a list, we place the instances that are more likely to be good substitutes at the front of the list. A second kind of structure is to add constraints to each instance on the list so that when we consider an instance, we look at the constraints on the instance and see if the current problem satisfies the constraints. An example of a constraint is "use a crowbar if the nail is very large." Finally, it might be useful to jump several categories up the hierarchy to try to find a substitute. If none of the instances in the category "nail extracting tools" worked, we look for possible substitutes in the categories of which "nail extracting tools" is an instance—for example "extracting tools" or "tools for nails."

Consider an example in the networking world. Suppose it has become routine to install any new software on a server named doc. We can express this general solution as follows, where the variable A has been instantiated by the names of several software applications:

Install the application A on the server named doc.

But now suppose that a new application requires 360K of disk space, and doc has considerably less than 360K of available disk space. The obvious solution that comes to mind is to substitute the piece of the solution "on the server named doc" with another piece—perhaps "on the server named anakana" or "on the server named coffee."

Some network administrators keep this sort of information in their heads, or they can poke around the network and come up with a workable solution fairly quickly. Other administrators, however, keep a log of the resources of the network and appeal to the log when these sorts of problems arise. A log of network resources is a good place to find substitutes for pieces of an old unworkable solution. In the example, we look through the log to find a server that has at least 360K of available disk space. The extra burden of this approach is the maintenance required to keep information about network resources current.

Another place that we can look for a new piece of the solution is in another case. For example, we can imagine several cases in the library that have the following structure:

Install the application A on server named B.

If we have an unworkable solution for some value of B, then we look for a similar case that has a different value for B regardless of the value for A. With this approach, we have to have a relatively rich and mature case library.

It is possible that we can come up with creative solutions when we substitute several pieces in an unworkable solution and put them together to form a new solution. Consider a proposed solution with the following structure.

Install the application A on server named B with remote access C.

Instances of the category "remote access" include rlogin, rpc, telnet, or nfs. "Remote access" itself is an instance of "access," which includes "remote access" and "local access." Given these possible substitutes for the piece "remote access C" together with the possible substitutes for the piece "on server named B," some alternative solutions will be offered.

3.4.3 Parameterized Adaptation

Definition:

Parameterized adaptation is a case adaptation technique in which a formula for deriving a solution in a retrieved case is used to find a solution to an input case.

A third style of adaptation is parameterized adaptation. With this method, the system looks at the relation between the problem and the solution in an old case and then uses the relation as a means to derive a new solution for an outstanding problem. If the problem descriptions in the old and outstanding cases are identical, then this method reduces to null adaptation. Otherwise, the system tweaks the old solution with respect to the relation and the new problem description.

Parameterized adaptation is commonly used when the relevant variables in a case are numeric, and we know the method or formula by which a solution variable is derived from the problem variables. A domain in which we commonly use this method is cooking. If we know that a 10-pound turkey takes 2.5 hours to cook at 325 degrees, then we would guess that a 7-pound turkey takes about 1.75 hours to cook, where the formula is "15 minutes of cooking for every pound."

In the networking world, for example, a general rule of thumb for determining the appropriate bandwidth to allocate to a subnet backbone is "50% of the sum of bandwidths of each connection to the backbone." Thus, if a problem in an old case were described as "performance degradation for subnet1" and the solution were to adjust the bandwidth of the subnet backbone according to this rule of thumb, then the same rule is applied to propose solutions to similar problems. The problem variables (i.e., the number of connections to the backbone and the bandwidth of each) might vary for different instances of the problem, but the general rule for determining a solution is the same.

One of the difficulties of applying this method is that of knowing the function that relates problem variables and a solution variable. The current and safest approach is to handcraft the function and attach it to the case. When the system retrieves the case, it will also retrieve the function and apply the function to the values of the variables in the current problem.

A more ambitious task is to design the CBR system so that it learns such functions when they are not readily available. Under this approach, a learning algorithm is executed over a collection of cases for a particular kind of problem. The algorithm approximates a function that predicts solutions for similar future problems, based on the collective solutions of the past problems. There is research in this area using neural nets as a function approximater; however, more work and understanding is required before this approach is practically feasible.

3.4.4 Procedural Adaptation

Definition:

Procedural adaptation is a case adaptation technique in which a procedure for deriving a solution in a retrieved case is used to find a solution for a problem in an input case.

A fourth style of adaptation is procedural adaptation. For example, in network troubleshooting manuals, we can find procedures or flowcharts that tell us how to go about pinpointing the cause of a problem and the possible solutions. Furthermore, troubleshooters often devise their own procedures that they publicize to the users of the network when problems occur. These procedures can be considered as decision trees or checklists that guide our investigations into the causes and possible solutions to a problem. An example of a general procedure for a malfunctioning workstation is this:

Check all cable connections.
Determine whether changes have been made to hardware.
Determine whether changes have been made to software.
Determine whether changes have been made to network configuration.
Question the last user of the workstation.

An example of a specific procedure for a problem with ejecting a diskette from an external diskette drive on a UNIX workstation is below. This procedure is implemented easily in a UNIX script.

```
if problem = "eject problem"
then

    if whoami is not equal to root
    then switch to root and execute "umount" and execute
    "eject"

    else if current directory is disk drive
    then get out of current directory and execute "umount" and
    execute "eject"

    else if execute "eject -q" = "diskette not in drive"
    then report "there is not a diskette in the drive"
```

Note that procedures are not solutions to problems but instructions that may give rise to the cause of a problem and potential solutions. With procedural adaptation, the system offers (or possibly executes) the procedure that worked for the old case as a solution to a current problem. To use this method, it is necessary to have a repository for procedures and a way to map kinds of problems to procedures that most often uncovered a solution. One way to do this is to attach a successful procedure to a case. Another way is to attach a pointer to a procedure and to collect procedures in a separate place.

3.4.5 Critic-Based Adaptation

Definition:

Critic-based adaptation is a case adaptation technique in which a user (i.e., critic) looks at a proposed solution in a retrieved case and manually adapts the solution to fit an input case.

We consider critic-based adaptation last, because this method includes the methods already discussed. The difference is that in critic-based adaptation, the burden of adapting a proposed solution rests with the troubleshooter rather than with the CBR system.

Frequently, the system retrieves solutions that we know will not work. We also know why they will not work and how to make them work. In these situations, it would not be worth the trouble to take the steps to try to determine whether the CBR system could adapt the case automatically. Instead, we adapt the case ourselves and rest with the fact that the case has simply suggested a ballpark solution that we can make work with a little tweaking. We should remember, however, to document the way in which we adapted the case so that we can use it again.

To do this, we should use language that indicates what is different in the new problem description compared with the old problem description. The difference is what makes the old solution unworkable and the new solution workable.

This is illustrated with a simple example. Suppose a new problem is submitted to the CBR system where the problem is described as "device hung":

```
device_name = holly
device_type = workstation
problem = device_hung
solution = ?
```

Assume that the system retrieves the following case and proposes the solution "reboot_device":

```
device_name = anakana
device_type = workstation
problem = device_hung
solution = reboot_device
```

But now suppose that one knows that holly is a server and that it is impolite to reboot a server without notice, as the clients that are connected to the server will be unprepared for the interruption. Thus, two steps are added to the proposed solution as follows:

```
Broadcast to clients of device "Device going down in fifteen
minutes to reboot"
Wait 15 minutes
Reboot_device
```

Suppose this solution is entered as a new case in the case library. Then the next time a problem of the type "device_hung" occurs, the system looks at the role of the device in the network before it recommends a solution. If the role of the device is "server," the system retrieves the case that proposes the solution that broadcasts a message and then reboots the device. Otherwise, the system retrieves the case that proposes "reboot device."

We have already shown several ways to incorporate this new bit of information into a CBR system. One way is to write a procedure that makes the distinction explicit and to provide a pointer to the procedure as the value for "solution" in the case. For example:

```
if device_role = client
then solution = reboot_device

if device_role = server
then solution =   broadcast_message_to_clients
                  wait
                  reboot_device
```

Another way is to file a new case that shows why the old solution was unworkable:

```
device = holly
device_type = workstation
problem = device_hung
solution = ?
   proposed solution:    reboot_device
   solution status: no_good
   reason if no good: device_role = server
```

A final option is to include the slot "device_role" in the problem description:

```
device = holly
device_type = workstation
device_role = server
problem = device_hung
solution = broadcast_message_to_clients
          wait
          reboot_device
```

3.5 CASE EXECUTION

In the preceding sections we looked at a number of ways in which a CBR system can retrieve a ballpark solution for a current problem and several ways to adapt a ballpark solution to fit a specific problem. Case retrieval and case adaptation were likened to the commonsense notions of appealing to experience and applying experience. The final step in the CBR process is to understand the notion of getting experience. Getting experience is construed as (a) executing a proposed solution and evaluating the results, and (b) embedding and organizing an experience with a proposed solution in the case library.

In this section, we make a distinction among three modes of executing a proposed solution:

- Manual execution;
- Unsupervised execution;
- Supervised execution.

In the following section, we examine the step of organizing the experiences with solutions in a case library.

3.5.1 Manual Execution

Definition:

With manual execution, the user of the system interprets a solution and decides whether to execute it.

Most current CBR systems only suggest good solutions to problems. Given the experiences in the case library and depending upon the retrieval and adaptation methods used, they offer their best shot at a solution. In manual mode, the user interprets the proposed solution and may choose to execute it. For example, solutions that are expressed in ordinary English have to be interpreted and executed manually by a user. On the other hand, solutions that are expressed in computer programs may be executed without user intervention, although it is risky to give CBR systems this much responsibility.

3.5.2 Unsupervised Execution

Definition:

With unsupervised execution, a proposed solution is executed automatically without user intervention or control.

It is conceivable that in some domains, a CBR system could take the extra step to execute the solution that it proposes. Where this is possible, we have a closed loop of problem solving without user intervention—a problem is submitted to the system, a similar case is retrieved and adapted in order to find a solution, the solution is executed by the system, and the results are embedded in the case library.

As an example of where this is feasible, consider the problem "pc floppy will not eject from UNIX disk drive." The simple UNIX script shown in Figure 3.8 can resolve this problem most of the time. If the script were named fix_pc_floppy_eject_problem, and the solution slot in the retrieved case held the name of the script, then the CBR system executes the script on the node that is experiencing the problem. Assuming that the CBR system has the proper permissions, it copies the script to the node, runs a remote shell command to execute the script, reports the results to the submitter, and deletes the script.

3.5.3 Supervised Execution

Definition:

With supervised execution, a proposed solution is executed automatically under the control of the user.

```
message='eject'
case "$message" in
 "floppy will not eject")
    if test 'whoami' != "root"
    then
       echo "You must be root to eject floppy. Attempting…"
       'su root'
       'umount /pcfs'
       'eject'
    elsif test 'pwd' = "/pcfs"
    then
       echo "You cannot eject the floppy while in directory /pcfs. Attempting…
       'cd /'
       'umount /pcfs'
       'eject'
    fi ;;
 "diskette NOT in drive")
    echo "A diskette is not in the floppy drive" ;;
 *)
    echo "Cannot fix the problem. Call MIS" ;;
esac
```

Figure 3.8 A script for the problem "pc floppy will not eject for UNIX disk drive."

An intermediate mode of execution is to consider a user as an over-seer that has the right to permit or forbid the execution of a proposed solution. The idea here is that the CBR system says, "Given my experience, I believe I can fix the problem by doing the following: … Shall I proceed?" and the overseer allows or disallows the execution of the proposal.

3.6 ORGANIZING CASES IN A CASE LIBRARY

We have examined several ways to express an experience with a proposed solution in a case. If the experience turns out to be an adaptation of an old solution, the system records the differences between the old problem and the new problem, showing why the old solution does not work and why the adapted solution works. The CBR retrieval mechanism exploits these differences when similar problems recur.

The question of how we embed these experiences and organize them in a case library is a different matter. So far, we have assumed that cases are embedded in a case library sequentially, much like an unordered stack of forms. If we need to retrieve a useful form, we sift through the stack one by one until we find a good match. If we reflect on how our own experiences are organized in our minds, however, we would eventually come to the conclusion that they are structured in quite complex ways.

In this section, we look at four organizational principles that have been used to structure a case library in CBR systems:

- Sequential memory;
- Hierarchical memory;
- Meshed memory;
- Master cases.

From a cognitive science perspective, these organizational principles can be considered at best as approximations of how our minds organize experiences. From an engineering perspective, however, they have shown to be useful for practical applications.

3.6.1 Sequential Memory

Definition:

A sequential memory is a structure in which the elements of the structure are organized as a list.

The most common structure of a case library is sequential memory. When we enter a case into a library, we simply put it at the end (or beginning) of a list. Although the algorithm that places the new case in the list is straightforward, the retrieval algorithm that searches the list can become overburdened. This is especially true when the number of cases is very large, and there is a matching routine that compares each slot in an input case with the slots in the cases stored in the list. Recall that we have argued for different reasons that we do not want to use such a strategy anyway, because not all of the slots in a case will be pertinent to the problem at hand.

There are several ways in which we can enjoy the simplicity of a sequential memory and lessen the burden on the retrieval algorithm. For example, when retrieving a case from the case library the first thing we want to do is look for those cases that exhibit the same problem. If the language that we construct for describing types of problems is predefined, then it is an easy matter in the beginning to collect those cases in a buffer whose problem type matches the current problem type. Then, we continue to prune those cases by performing the matching routine with respect to increasing refinements of the problem.

For example, a problem might be refined as follows:

communications error: access error : rlogin error

The retrieval algorithm puts pointers on all cases that match with "communication error" in an initial buffer; then performs the matching routine on these cases with respect to "access error" and puts those pointers in a second buffer; and finally matches on these cases with respect to "rlogin error." A useful feature of this strategy is that we can back up to the previous buffer if good solutions in a current buffer are not found.

This searching strategy is also useful if we use relevance rules to guide the search. Recall that the form of a relevance rule is the following:

If the problem is of type P_1, then match on subset S_1 of S.

Our matching routine considers the subset S_1 of slots when retrieving potentially useful cases. Those cases that matched on the whole of S_1 are placed in a buffer for first consideration, and those that matched on portions of S_1 are placed in another buffer for secondary consideration.

Another way in which we can enjoy the simplicity of a sequential memory is to execute a cleanup mechanism periodically over the case library. If we have a case that records a routine problem and solution, then it is likely that a sequential case library will eventually contain many duplicate cases that are scattered throughout the library. The cleanup mechanism deletes any duplicate cases. If it is important to keep a record of all cases, the cleanup mechanism stores duplicate cases in a secondary library.

3.6.2 Hierarchical Memory

Definition:

A hierarchical memory is a structure whose elements are organized as types or subtypes and where each element has, at most, one supertype.

In the preceding section, the burden of case organization was placed on the case retrieval mechanism. The mechanism imposed an artificial structure on a sequential library by dividing the case library at retrieval time.

From a cognitive science perspective, however, it is doubtful that our experiences are organized sequentially. A prevalent hypothesis about the structure of experiences is that they are organized hierarchically. At the top of the hierarchy is the broad label "an experience." Particular kinds of experiences reside on a level beneath the top, for example "an experience with people" and "an experience with a mechanical device." In turn, each of these subcategories are divided into more specific kinds of experiences.

Given a completed experience, the system finds the most specific category in the hierarchy of which the experience is an instance. Then the system embeds the experience within this category of the hierarchy.

The importance of this hypothesis for building CBR systems is that we can place the burden of case organization on the mechanism that embeds cases in the library. Under this approach, the case library is organized at embed time rather than at retrieval time. The trade-off is that there is more of a burden on the embedding mechanism, albeit less of a burden on the retrieval mechanism.

From a practical perspective, the simple way to structure a case library at embed time is to group the cases with respect to problem types. When an experience is ready to be entered in the library, we find the place in the hierarchy that best fits the case. For example, Figure 3.9 shows a partial hierarchy for networking problems.

The embedding mechanism searches the hierarchy to find the most specific category of problems to which a current problem belongs. For example, the following description of a problem is placed in the bottom-most left-most category.

problem: communications : access error: rlogin error

When a similar problem recurs, the retrieval mechanism searches the hierarchy instead of examining each case individually. If the number of cases becomes increasingly large, the burden on the retrieval algorithm increases at least linearly with respect to the number of cases.

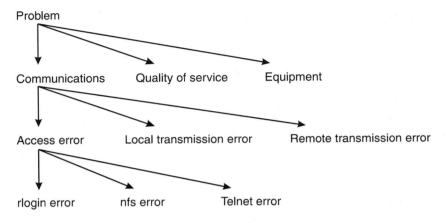

Figure 3.9 A hierarchical memory for representing network problems.

3.6.3 Meshed Memory

Definition:

A meshed memory is a structure in which the elements of the structure are organized as types, subtypes, etc. and where each element can have multiple supertypes.

An assumption of a strictly hierarchical memory is that any one category has at most one supercategory. Figure 3.9, for example, shows that the problem "rlogin error" is a type of "access error." In a meshed memory, this assumption is lifted. We allow a category of experience to have multiple supercategories. We might allow the problem "rlogin error" to be considered also as a kind of "remote transmission error." Figure 3.10 illustrates a simple example.

The advantage of a meshed memory is that it allows us to open up more potential solutions to problems that we encounter. Suppose we collect the cases in memory that have the following description:

problem: communications : access error: rlogin error

If none of these cases has a viable solution, the system looks for other solutions in cases that have the description:

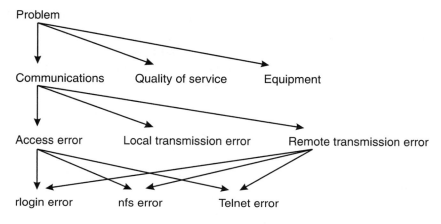

Figure 3.10   A meshed memory for representing networking problems.

problem: communications : remote transmission error: rlogin error

Intuitively, the concept of a meshed memory is closer to the way our own experiences are organized in our minds. In all but simple domains, however, it is difficult to conceive of the appropriate structure of a meshed memory. Furthermore, meshed memory imposes a greater burden on the retrieval algorithm. The retrieval algorithm has to be smart enough to know how to backtrack to alternative supercategories if a solution is not forthcoming in the first selected category.

3.6.4   Master Cases

Definition:

A master case is a case structure that subsumes experiences with similar individual cases. When an experience with a case is ready to be submitted to the case library, it is incorporated into a master case (if one exists).

The concept of a library of master cases is simple and fairly easy to implement. The idea is that when we find an alternative solution to a solution that is already recorded in the library, we append the solution to the existing case rather than create a new case. Thus, for each problem type, there will be just one case that accumulates alternative solutions over time. Each solution in the case will have the form "if X then the solution is Y," where X holds special considerations regarding the problem, and Y holds the solution if X is true.

We have already seen the beginnings of a master case for the problem "device_hung." Some additional alternative solutions to this problem might be recorded in a master case as follows:

```
if problem = device_hung
then

if console_message = server_not_responding
then solution = do_nothing

if operable_window_exists = yes AND hanging_process_known = yes
then solution = kill_hanging_process

if operable_window_exists = no AND device_function = client
then solution = reboot_device

if operable_window_exists = no AND device_function = server
then solution = broadcast_message
                wait
                reboot_device
```

Note that a master case has the look of a mini ES that covers possible solutions for a problem type. Therefore, it is arguable that the problems with ESs in general apply equally to master cases. If we were to execute a master case in the same manner in which we execute an ES, we would have the problem of arbitration when the system suggests multiple solutions, and we would have the problem of frustration if no solution were forthcoming.

The CBR approach to problem solving, however, is not as ambitious in this regard as the ES approach. Coming up with an ultimate correct solution to a problem is not as important as devising some ballpark solutions to a problem and letting the user do the rest. For example, it would be useful to propose a list of alternative solutions to a problem and the conditions under which the solutions were known to work. An additional feature of the CBR paradigm that is not shared by the ES paradigm is that it is a simple matter to append newly found solutions to a master case for consideration in instances of future similar problems.

3.7 SUMMARY

This chapter has presented the CBR framework for representing and solving problems. The goals of the chapter were to place the CBR framework in perspective, to look at various styles of designing CBR systems, and to suggest some of the kinds of problems in the networking world for which CBR systems are useful.

The main requirement of a troubleshooting tool in the networking domain is that it exhibit some degree of learning, adaptability, and evolution as the network operation evolves. This chapter showed how the CBR approach to problem solving satisfies this requirement.

Early in the chapter, Figure 3.1 showed the basic structure of a CBR system. In the networking domain, a CBR system includes a case library that documents prior experiences with network problems. When a new problem occurs, the system retrieves a similar problem in the library and tries to adapt the solution in the old problem in order to propose a solution for the new problem. The experience with the proposed solution—whether it is successful or not and why—is entered back into the library.

The goals of the CBR system are to grow at roughly the same pace as the network operation, to be able to at least offer ballpark solutions to network problems based on past experiences, and to require little maintenance.

The body of the chapter comprised discussions of the different ways in which CBR system builders have approached the tasks of designing a case, the retrieval algorithms, adaptation algorithms, execution procedures, and the organizational structure of a case library. Figure 3.11 sums up the available options. These options should be considered as a CBR toolbox, where different applications require different sets of the tools. In the following chapter, we will look at some concrete applications of CBR.

3.8 FURTHER READING

For further reading in the general CBR framework of problem solving, two good books are Kolodner's *Case-Based Reasoning* and Riesbeck's *Inside Case-Based Reasoning*. Kolodner, Simpson, and Sycara-Cyranski's paper "A Process of Case-Based Reasoning in Problem Solving" is considered the seminal work on CBR. For shorter, general discussions of CBR refer to the following papers: Barletta's "An Introduction to Case-Based

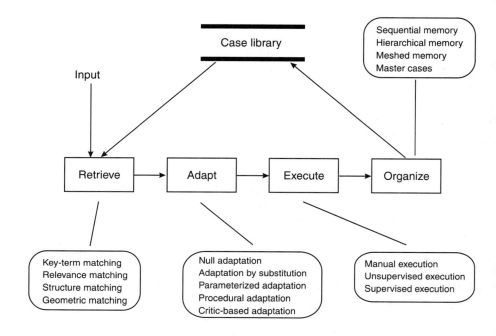

Figure 3.11 Design options for CBR systems.

Reasoning," Kolodner's "Improving Human Decision-Making Through Case-Based Decision Aiding," and Slade's "Case-Based Reasoning: A Research Paradigm."

For further reading on William Bain's work on CBR and legal procedure, see "Case-Based Reasoning: A Computer Model of Subjective Assessment" or the shorter "A Case-Based Reasoning System for Subjective Assessment."

The method of case retrieval based on relevance matching was inspired by Davies and Russell's paper "A Logical Approach to Reasoning by Analogy." In the paper, relevance rules are referred to as determination rules. For work on deriving a set of relevance rules automatically from an existing case library, see Lewis and Sycamore's "Learning Index Rules and Adaptation Functions for a Communications Network Fault Resolution System" in *Applications of Neural Networks to Telecommunications* and Rissland and Skalak's "Combining Case-Based and Rule-Based Reasoning: A Heuristic Approach."

The method of case retrieval based on structure matching was inspired by Gentner's paper "Structure-Mapping: A Theoretical Framework for Analogy." The method of case retrieval based on geometric distance was inspired by Stottler, Henke, and King's paper "Rapid Retrieval Algorithms for Case-Based Reasoning."

For further reading on the concept of a master case, see Dreo and Valta's "Using Master Tickets as a Storage for Problem Solving Expertise" in *Integrated Network Management IV*.

For further reading on fuzzy matching and fuzzy inference, see Lewis and Dreo's "Extending Trouble Ticket Systems to Fault Diagnostics."

Select Bibliography

Bain, W., "Case-Based Reasoning: A Computer Model of Subjective Assessment," Ph.D. diss., Department of Computer Science, Yale University, 1986.

Bain, W., "A Case-Based Reasoning System for Subjective Assessment," in *Proceedings of the National Conference on Artificial Intelligence*, Los Altos, CA: Morgan Kaufmann Publishers, 1986.

Barletta, R., "An Introduction to Case-Based Reasoning," *AI Expert*, August 1991.

Davies, T., and S. Russell, "A Logical Approach to Reasoning by Analogy," in *Proceedings of the International Joint Conference on Artificial Intelligence*, Los Altos, CA: Morgan Kaufmann, 1987.

Gentner, D., "Structure-Mapping: A Theoretical Framework for Analogy," *Cognitive Science*, 7, 1983.

Dreo, G., and R. Valta, "Using Master Tickets as a Storage for Problem Solving Expertise," in *Integrated Network Management IV*, W. Zimmer and D. Zuckerman (eds.), Amsterdam: North Holland/Elsevier Science Publishers, 1995.

Kolodner, J. L., "Improving Human Decision-Making Through Case-Based Decision Aiding," *AI Magazine*, Vol. 12, No. 2, 1991.

Kolodner, J. L., *Case-Based Reasoning*, San Mateo, CA: Morgan Kaufmann, 1993.

Kolodner, J. L., R. Simpson, and K. A. Sycara-Cyranski, "Process of Case-Based Reasoning in Problem Solving," in *Proceedings of the International Joint Conference on Artificial Intelligence*, Los Altos, CA: Morgan Kaufmann, 1985.

Lewis, L., and G. Dreo, "Extending Trouble Ticket Systems to Fault Diagnostics," *IEEE Network*, Vol. 7, No. 6, 1993.

Lewis, L., and S. Sycamore, "Learning Index Rules and Adaptation Functions for a Communications Network Fault Resolution System," in *Applications of Neural Networks to Telecommunications*, J. Alspector, R. Goodman, and T. X. Brown (eds.), Northvale, NJ: Lawrence Erlbaum Associates, 1993.

Riesbeck, C. K., and R. S. Schank, *Inside Case-Based Reasoning*, Northvale, NJ: Lawrence Erlbaum Associates, 1989.

Rissland, E., and D. Skalak, "Combining Case-Based and Rule-Based Reasoning: A Heuristic Approach," in *Proceedings of the International Joint Conference on Artificial Intelligence,* Los Altos, CA: Morgan Kaufmann, 1989.

Slade, S., "Case-Based Reasoning: A Research Paradigm," *AI Magazine,* Vol. 12, No. 1, 1991.

Stottler, R., A. Henke, and J. King, "Rapid Retrieval Algorithms for Case-Based Reasoning," in *Proceedings of the International Joint Conference on Artificial Intelligence,* Los Altos, CA: Morgan Kaufmann, 1989.

Examples of Successful Cased-Based Reasoning Applications

4

In Chapter 4:

❏ Prism: A Case-Based Telex Classifier
❏ Canasta: A Crash Analysis Troubleshooting Assistant
❏ Cebrum: Case-Based Reasoning in Large-Scale Manufacturing
❏ Smart: A Case-Based Reasoning Call Support System

CBR became recognized as a viable framework for understanding and implementing problem-solving tasks in the early 1980s. As with most new theories, CBR applications were research-oriented in the beginning. Practical, deployed applications began to emerge by the late 1980s.

This chapter presents case studies of four deployed applications that use CBR methods described in the previous chapter. The purpose of this chapter is to gain secondhand experience that we can carry over to the network management domain. For each application, we will be interested in the task that was achieved, the design of the CBR system that was constructed for the task, the way in which the application was integrated into existing operations, and the payoff from the CBR approach compared to previous approaches. In addition, we will discuss the applications' implications for network management.

The applications were selected from the *Innovative Applications of Artificial Intelligence (IAAI)* book series, which was initiated in 1989 by the American Association of Artificial Intelligence. (See Section 4.6 for details.) The requirements for having an application included in the IAAI books are as follows:

1. The application is in use in an operational environment.
2. The application demonstrates payoff—in terms of competitive advantage gained, enhanced quality, or enhanced productivity.
3. The application compares favorably with respect to payoff vs. the cost of development.
4. The application's payoff justifies the cost of its maintenance.
5. The application demonstrates the mechanisms by which validation of performance is achieved.

These are stringent but necessary requirements for a successful application. The examples will demonstrate the kinds of results that can be achieved with CBR and provide us with pointers for designing and building our own CBR applications for problems in network management.

4.1 PRISM: A CASE-BASED TELEX CLASSIFIER

4.1.1 Background

International banks receive from hundreds to thousands of telexes per day via an interbank communications network. A telex is a brief, truncated message sent from a correspondent bank. The message emits from a printer at the receiving bank in close to real time. An example of a typical bank telex is shown in Figure 4.1.

The two main operations performed on an incoming telex are (a) to classify the message, and (b) to route the message to the proper destination in the bank. If these operations are not performed accurately and in a timely manner, they can increase bank expenses considerably.

The task of telex classification and routing was analyzed by Cognitive Systems, Inc. (CSI). CSI found that there are little more than a hundred possible telex classifications. It also determined that, on average, the existing methods of performing the task have a receipt-to-destination time of one minute and an accuracy rate of 75%.

```
FDB857
0221 02/16 001 730 DHS281 DGR160
433
TO CHAMANBANK NEWYORK
FM KOREA EXCHANGE BANK SEOUL FEB 16 1988
TEST 917626 FOR USD 180,626.-
PLS PAY USD 180,626,94 THA DAIWA BANK NEWYORK ATTN LOAD DEPT
WITHOUT DEDUCTING ANY COMMISSIONS VALUE FEB 16, 1988 VIA CHIPS
9
UID 013498 MENTIONING OUR IDENTIFICATION NBR 305078 AND
FOLLOWING RS N
R
G
SSLSDQAUG
LGSM STP
OUR L/C NBR PRIN INTEREST AMT THEIR REF
M0645702EUO0064 USD 54,625 USD4,059.64 356374
M0645702EUO0057 USD 23,862.50 USD1,773.42 356375
M0645702EUO0040 USD 15,208.71 USD1,130.29 356377
M0645702EUO0025 USD 49,925.00 USD3,710.34 4 0/
24
M0645612EUO0089 USD 21,856.25 USD1,624.32 356385
/.0645707EU000071 - USD2,851.47 356576
BEST RGDS
KOEXBANK MULL AEDONG BR
KEBSEL
135171 0729 160288 02700547 0726
02760421 857
NNNN
```

Figure 4.1 An example of a typical bank telex.

The kinds of messages in a telex include posting a posit or debit to an account, verifying a posit or debit, updating the status of an account, and transferring a sum of money from one account to another. However, the task of determining the classification is not trivial and requires special training. For example, the telex in Figure 4.1 requires a credit authorization to pay or accept, although to an outsider this is difficult to see.

Some banks have operators who perform these tasks manually. An operator tears off a telex when it emits from the printer, examines and classifies it, determines where it should go, and puts the copy in interoffice mail to be hand delivered to the proper destination.

An alternative method is to have operators who disperse telexes by computer, thereby eliminating the need for paper documentation. Each incoming telex is placed on a queue. The operator pops a telex off the queue, displays it, determines where it should go, attaches a routing code, and sends it off to the proper destination over the bank's internal network.

In either case, the task of telex classification and routing translates into operational overhead. Clearly, the computerized method is an improvement over the manual method, but both methods are expensive and create problems for the bank. For example:

1. Training operators to perform the task is expensive.
2. If the computerized method is used, there is the additional expense of training the operator to use a computer.
3. Operator turnover is high, adding extra expenses for recruiting and training new operators.
4. Telexes are transmitted 24 hours a day, whereas operators generally work day shifts. Thus, there are backlogs of telexes to be routed every morning. In addition, printers and networks sometimes go down, or telexes can come in surges. These situations increase the likelihood of operator stress and, therefore, increase the probability that operators will misroute or lose telexes.

Initially, CSI developed a special-purpose rule-based ES to perform the task. The ES showed an improvement over the existing method with an average receipt-to-destination time of 44 seconds and an average accuracy rate of 76%. One ES was put into operation at Chase Manhattan Bank in 1988 and another one at Societe Generale de Banque in 1989.

Over time, however, CSI discovered that the maintenance of the ESs added an extra, unacceptable expense to the solution. Whenever the rule base was modified to handle a new classification or a misclassification, the fix inevitably interfered with the part of the system that had previously worked correctly. The time required to trace and fix the new problems was prohibitive. In addition, the work required to customize an ES for a new installation was prohibitive, requiring about a year for one person.

CSI later developed a CBR system that exhibited an average receipt-to-destination time of 30 seconds and an average accuracy rate of 90%. The system, named Prism, was put into operation at Chase Manhattan Bank's Letter of Credit Department in October 1989 and has been cited as the first commercially deployed CBR system. In addition to performing

better than prior approaches, Prism required less customization for installation at new sites and needed considerably less maintenance. Furthermore, the system decreased the staffing requirement from five people to
three people or less.

4.1.2 The Structure of Prism

Prism consists of three modules—a lexical pattern matcher, a CBR module, and a rule-based router. Figure 4.2 shows Prism's architecture.

An incoming telex is passed first to a lexical pattern matcher that
translates the telex into a set of symbols. The pattern matcher is pro-
grammed to translate misspellings, synonyms, and abbreviations into a
consistent symbolic representation. For example, Figure 4.3 shows the
symbols that represent the telex in Figure 4.1.

Given a symbolic representation of the new telex, the CBR module
retrieves a case from its case library that is most similar to the input case
and returns the classification of the retrieved case. Finally, this classification is passed to a rule-based router that correlates classifications with
destinations in the bank.

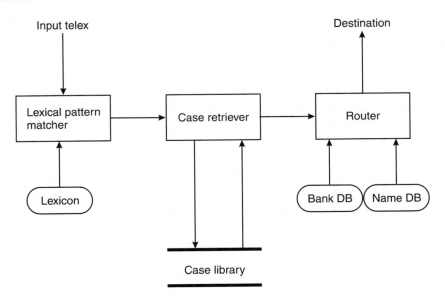

Figure 4.2 The structure of Prism.

```
TEST
USD
PAY-DEFINITE
LOAN
DEPARTMENT
WITHOUT
BANK-CHARGES
VALUE-DATE
THROUGH
MOP-CHIPS
MOP-ABA
MOP-UIC
LCRED-INFO
INTEREST-PAY
REF-NUMBER
```

Figure 4.3   A symbolic representation of the telex in Figure 4.1.

4.1.3   Features: A CBR Case-Building Tool

CSI developed a case-building tool that allows experts to construct a symbolic language by which to represent input cases and the cases in the library. A bank expert examines a telex and associates symbols with the content of the telex. The tool also allows the expert to build formulas that associate a collection of components in the telex with an abstract symbol in the language. This process provides the system with a basic lexicon for case representation. The lexical pattern matcher scans a new telex and uses the lexicon to represent the case in a structure that is more manageable than a raw telex.

An innovative feature of Prism is the method used to develop the case library and the mechanism used to retrieve similar cases. The library consists of more than 9,600 cases of previously classified telexes. However, the library, rather than being a sequential memory as one might expect, is a hierarchical memory.

The hierarchical memory is developed as follows: Each telex in the library is first run through the lexical pattern matcher as described above. Then the domain expert attaches a symbol to the case that represents its classification. So far, then, the case library is a sequential, symbolic list of classified telexes. The next step is to run the sequential list through an induction algorithm that transforms the list into a decision tree. The decision tree represents the entire case library with the intermediate nodes of the tree representing relevant features and the leaf nodes representing classifications. When a new symbolic telex is submitted to the system the decision tree is executed against the symbols describing the telex, retrieving a similar case and a classification.

The algorithm that transforms the case base into a decision tree belongs to the class of information theoretic (IT) algorithms. Let us illustrate the algorithm with a simple, albeit abstract, example. We will show a sample input (a sequential list of cases) and the resulting output (a decision tree). To find out how the algorithm works, refer to the citations in Section 4.6.

See Table 4.1. The slots that describe a case are $S1$, $S2$, $S3$, and C, where $S1$ takes on symbolic values x, y, z; $S2$ takes on values t, u, v; and $S3$ takes on values a, b, c. C is a classification slot (i.e., a solution slot) that takes on values d, e, and f. Table 4.1 shows a sample case library consisting of 12 cases.

Table 4.1
A Sequential Case Library in the Abstract

Case	S1	S2	S3	C
1	x	t	a	d
2	x	u	a	d
3	x	u	b	e
4	y	t	a	d
5	y	t	b	e
6	y	u	b	e
7	x	v	a	d
8	y	v	b	e
9	x	u	c	f
10	y	t	c	f
11	z	u	b	e
12	z	t	a	d

A decision tree is derived from Table 4.1, as shown in Figure 4.4. The tree shows the relevance of each slot to the prediction of the classification. It is important to note that in the example, only slot S3 is needed to determine the classification of a case. Thus, when a new case is submitted to the system, the retrieval algorithm has only to consider the value of S3. One can see this by inspecting the correlation between the columns S3 and C in Table 4.1.

Now, if the decision tree classifies a new case incorrectly, one adds the new case and its correct classification to the original list and runs the transformation algorithm again, thus producing a new decision tree. In this way, the decision tree is updated.

It will be rare that a sequential case base can be transformed into a decision tree that is as clean and elegant as the tree in Figure 4.4. The worst scenario is that the tree will be bushy, where there extend distinct paths from the root node to a leaf node for each case in the case base.

Note that an IT algorithm is a method of deriving a set of relevance rules from an existing case base automatically (see Section 3.3.2). The algorithm selects those slots that are relevant to a set of classifications and discards the rest.

A problem to worry about in using this method is the possibility of mistaking a coincidental correlation between two slots for a causal correlation. For example, a slot in the problem description may be thought to be causally relevant to a solution slot when really there is only a happy coincidence. In the example above, it would be wise to question whether the correlation between S3 and C is coincidental or real. Nonetheless, the method can be powerful if the domain language is designed carefully and the development of the case base is controlled properly.

In sum, Prism combines several methods with CBR: a lexical pattern matcher that transforms telexes into a consistent case representation, an

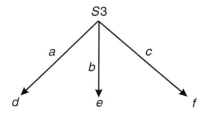

Figure 4.4 A decision tree for the case base in Table 4.1.

IT algorithm that transforms a sequential case base into a decision tree, and a rule-based system that retrieves bank destinations for particular classifications.

4.1.4 Implications for Network Management

On the surface, the classification of bank telexes appears to be similar to the classification of network events. Thus, we might have the makings of an approach to a difficult problem in network management—the task of translating raw network events to network problems and routing them to the responsible parties. We will work out a first approximation to this idea below.

Most network management platforms include an event logger that logs all the network events to a log. Events can be messages that indicate the powering up or powering down of a device, the loss of contact with a device, the overstepping of a threshold of a device parameter, and other happenings on the network. Alternatively, an event logger may be configured to report the values of the parameters for a single device in time increments. An event log for a router might include packet rate, packet source and destination, network load, soft error rate, and hard error rate. For example, Figure 4.5 shows a small portion of a list of events that is logged by a network event logger.

Note that the representation of a network event shares some of the properties of a bank telex. There are a lot of them (comparatively more than telexes); they are unstructured or at best semistructured; they can be conceived of as messages that might require an action by a network administrator; and although there exist a few network experts who can translate an event log into useful information, the task is difficult.

A current, useful way to understand the behavior of a network is to run the event log through a report generator that produces graphical reports of network behavior over time. For example, one can see the amount of traffic over different segments in the network or see the proportion of uptime vs. downtime of individual devices on the network.

On the other hand, the event log can be analyzed with an eye toward identifying problems on the network, identifying security loopholes, planning for network growth, or distributing costs to network users. While the production of graphical reports is relatively mature, the way to achieve these latter goals is less understood.

```
03/05/94 12:19:20, 10328, 402b85, 1009b INT 479, 10098 INT 3695211, 1009f
INT 3694678, 100a8 INT 0, 1009d INT 0

03/05/94 12:29:21, 10328, 402b85, 1009b INT 479, 10098 INT 3695310, 1009f
INT 3694777, 100a8 INT 0, 1009d INT 0

03/05/94 12:49:32, 10328, 402b85, 1009b INT 479, 10098 INT 3695488, 1009f
INT 3694955, 100a8 INT 0, 1009d INT 0

03/05/94 12:58:52, 10328, 402b85, 1009b INT 479, 10098 INT 3695561, 1009f
INT 3695028, 100a8 INT 0, 1009d INT 0

03/05/94 13:09:00, 10328, 402b85, 1009b INT 479, 10098 INT 3695640, 1009f
INT 3695107, 100a8 INT 0, 1009d INT 0

03/05/94 13:19:59, 10328, 402b85, 1009b INT 479, 10098 INT 3695731, 1009f
INT 3695198, 100a8 INT 0, 1009d INT 0
```

Figure 4.5 Sample printout of a network event logger.

A viable approach toward understanding how to achieve these goals is to cast them in a Prism-like CBR framework. For example, if the task were to identify problems on the network and route them to the attention of a network troubleshooter, one starts by developing a symbolic language for representing network events and network problems. A special-purpose event parser would be used to symbolize incoming events. To develop the case library, a domain expert associates a symbolic event with a known problem type. Finally, the library is transformed into a decision tree that is executed against incoming events.

Although this idea sounds promising, the network event classification problem will probably be more difficult than the telex classification problem. For one thing, there are considerably more events to classify. Moreover, only a small portion of the events are significant. Finally, a particular pattern of events might indicate a network problem, although any one of the events in isolation would not indicate the problem. It will be hard to define the various patterns of events that can be mapped to specific kinds of network problems, and, given the abundance of events available for inspection, it will be hard to discern the patterns in the log. Nonetheless, the idea shows promise and is worthy of pursuit.

4.2 CANASTA: A CRASH ANALYSIS TROUBLE-SHOOTING ASSISTANT

4.2.1 Background

The domain for this application is the support of computer systems. Given the complexity of computer hardware, operating systems, and software applications, it is inevitable that some normal operation of these components in tandem will cause a computer system to crash. Even though each component of the system might have passed a stringent quality assurance test, users of the system seem to find a path through the system that crashes the system. Customers are not happy with system crashes. They report them to computer manufacturers and expect the problem to be fixed.

In a typical manufacturing organization, product support engineers are the initial contacts with customers who have problems with the system. If support engineers cannot fix the problem, it is handed over to development engineers. Clearly, a manufacturing organization can operate more efficiently if customer problems can be handled by frontline support engineers, allowing the developers to spend more time designing and building new products and enhancements to current products. Furthermore, the timeliness of finding a good fix translates into less operational overhead for both the manufacturer and the customer and contributes to customer satisfaction.

For these reasons, support engineers at Digital Equipment Corporation (DEC) studied their existing method of resolving VAX computer crashes at customer sites. They found a typical scenario to be as follows. When a customer reported the crash problem, the support engineer remotely connected to the customer's machine to read the system's crash dump file. The engineer executed scripts against the file that returned key information regarding the nature of the crash. The engineer then used this information to scan textual databases that documented previous experiences with crash problems and their solutions. The usual method to do this had been to perform a key-word search over the databases looking for similar problems.

If this method did not uncover a similar experience and a fix, the next phase was to look at the stack of procedure calls made prior to the point at which the system crashed. This approach required considerably more expertise on the part of the support engineer—knowledge of the assembly language and of how the operating system works. Finally, if a solution to the crash problem was not forthcoming from this method, the problem was handed to a development engineer for further analysis. This last approach usually involved a continuing dialogue between the development engineer, the support engineer, and the customer until the problem was resolved.

A study at DEC showed that on the average it took about 30 minutes for a support engineer to scan textual databases in search of a similar problem, and even then solutions were not always found. For example, the engineer might fail to scan the particular database where a similar problem had been documented; the key-word search might fail to bring up a relevant document even though it existed; the key-word search might bring up an irrelevant document, thereby wasting time; and there might not be a document that described the problem, although the problem was known to some members of the organization.

DEC initiated a project to design and develop a system to streamline their method of dealing with VAX crashes. By the end of 1989, the system, named Canasta, was deployed at DEC support sites worldwide. It is estimated that the system saves the company over two million dollars each year. This amount is determined by the decrease in time required to process a customer complaint. Additional savings accrue from the more accurate solutions afforded by the system, thereby reducing the wasted time and expense in trying erroneous solutions.

4.2.2 The Structure of Canasta

Canasta is an integrated rule-based/case-based system. The structure of Canasta is shown in Figure 4.6. The system consists of five modules, described below. Each module corresponds roughly to each phase of the previous method of crash investigation.

Data Collection Module

This module corresponds to the phase of collecting information from the crash dump file at the customer site. The module is a rule-based system that controls the execution of data collection commands that are available in a VAX dump analyzer. Based on the results of prior commands, the system determines the appropriate commands to execute next. Each command scans the binary crash dump file and returns ASCII text that repre-

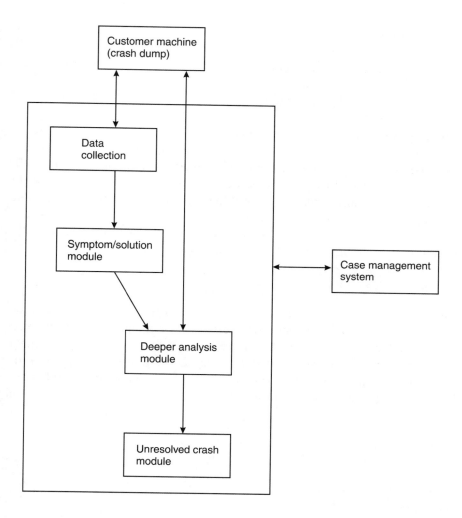

Figure 4.6   The architecture of Canasta.

sents a parameter of interest. The module collects up to 15 key parameters that are passed to the system/solution module.

Symptom/Solution Module

This module corresponds to the first line of attack taken by support engineers. The module is a rule-based system, where each rule has the following form:

If the symptoms are...

Then the problem description is...

And the technique for confirmation is...

And the solution or recommendation is...

The if part of a rule holds the symptoms that are matched against the parameters returned by the data collection module. The then part of the rule holds the likely cause of the problem, ways to confirm the cause, and how to repair the problem. An example of a rule is shown in Figure 4.7.

If no rule in the symptom/solution module suggests a fix for the problem, the user invokes the deeper analysis module.

Deeper Analysis Module

This module corresponds to the second line of attack taken by support engineers. The module consists of a collection of decision trees or instructions such as those that can be found in VAX troubleshooting manuals. They are encoded online so that the engineer can walk through the instructions, using information provided by the previous module and possibly gathering more information by running tests or further analyzing the

```
IF:
    VMS-Version      = one-of 5.0, 5.0-1, 5.0-2, 5.1, 5.1-1
    Bugcheck-Type    = INVEXCEPTN
    Module-of-Crash = ETDRIVER
    Module-Offset    = x1546
THEN:
 DESCRIPTION:
    The crash occurs due to a synchronization problem in the ETDRIVER
    while it is stepping down its list of UCB.
 TECHNIQUE FOR CONFIRMATION:
    Check to see if R5 = FFFFFFFF. This tells us that the UCB has already
    been severed.
 SOLUTION-RECOMMENDATION:
    1. You should first recommend the customer upgrade to VMS version
       V5.2 because this patch and 6 others are included in this major release.
    2. If the customer cannot upgrade then send them patch number 0055.
```

Figure 4.7 A sample rule in Canasta's symptom/solution module.

crash dump files. If a fix is not forthcoming from this exercise, the problem is passed to the unresolved crash module.

Unresolved Crash Module

This module is the repository for any system crash that is not repaired via the symptom/solution module or the deeper analysis module. Unresolved crashes are entered into a library and are then escalated to seasoned support engineers or to development engineers.

A useful feature of the module is a preprocessor that groups unresolved crashes into similarity sets and suggests possible underlying causes for each member in the set. The processor is guided by high-level, commonsense heuristics for grouping crashes and suggesting ballpark causes. Examples of two such heuristics are these:

Crashes occurring in the same software module with very similar offsets are most likely due to the same underlying cause.

If all the crashes in a set occurred in the same minor release of VMS, then the problem seems to be in that minor version of the VMS software (as opposed to hardware or the microcode).

For each similarity set, the processor develops a master case that contains the common symptoms among the cases in the similarity set, a probable cause, and pointers to each particular case. When a new problem crash is submitted to the module, the symptoms describing the new crash are matched with respect to the master cases. In this way, a user with a new crash problem can retrieve a master case that is similar to the problem and navigate to specific cases in order to examine comments or further progress.

Case Management System

This module records all experiences with crash dump problems in a common database. Each entry includes a slot for a status of *resolved* or *unresolved* and a slot showing the method that was successful in finding a solution to the problem—whether it was resolved by the symptom/solution module, the deeper analysis module, or by a human expert. We should note that the database serves as a way to update each of the modules with new or revised knowledge about crash dumps. For example, an engineer periodically retrieves all entries that are resolved by non-Canasta means and uses the resolution method to update the rule base of the symptom/solution module or to modify a decision tree in the deeper analysis module.

4.2.3 Features: The Multitiered Approach to Problem Solving

An innovative feature of Canasta is the multitiered approach to problem solving. The first tier attacks a crash problem with a rule-based system (the symptom/solution module), the second tier attacks the problem with decision trees (the deeper analysis module), and the third tier attacks the problem with CBR methods (the unresolved crash module). In addition, a lateral tier—the case management system—collects all experiences with crashes and uses critic-based adaptation to refine the knowledge in the other modules.

Note that Canasta is not a CBR system proper as described in Chapter 3. Nonetheless, we can see CBR techniques at work in several places in the overall system. In particular, the system to some extent alleviates the problem of knowledge acquisition in traditional rule-based systems. Although the multiple rule-based systems have to be updated manually as new experiences with VAX crashes occur, the process is guided by the case management module.

4.2.4 Implications for Network Management

Two lessons that can be gleaned from the success of Canasta are the following:

- CBR is not a panacea for all problem-solving tasks.
- In some applications, CBR techniques can be interleaved with other methods of problem solving.

One should keep these lessons in mind when considering applications of CBR for network management tasks. In Chapter 5, we will examine some of the problem-solving capabilities of current network management tools and see how we can complement the tools with CBR.

4.3 CEBRUM: CBR IN LARGE-SCALE MANUFACTURING

4.3.1 Background

The domain of this application is the manufacture of submarines. When we think of large-scale manufacturing, we should think of building a large, complex item over a long period of time as opposed to building a large number of comparatively smaller items over a shorter period of time. For example, it takes more than five years to build a submarine, whereas a single automobile plant builds thousands of automobiles in one year.

Large-scale manufacturing in the former sense gives rise to a costly problem—redundant engineering. Redundant engineering occurs when two or more similar problems are studied and resolved in relative isolation, without communication or sharing of information among the problem solvers.

The phenomenon of redundant engineering is ubiquitous and occurs to some extent in almost all domains. However, it is magnified in submarine manufacturing for the following reasons:

1. Different engineering departments are responsible for different sections of a submarine at different stages of development.
2. Because each section involves like components in fluid, electrical, structural, and mechanical systems, there is duplication in problem solving that occurs over these components.
3. There is the usual problem caused by turnover of personnel in each department.

The degree to which one can lessen redundant engineering in large-scale manufacturing is reflected in reduced overhead costs. For this reason, General Dynamics studied the problem in hopes of finding ways to alleviate the redundancy.

As an illustration, consider the following example. A standard valve is used in more than a thousand fluid systems in a submarine. Each valve is received, inspected, installed, tested, and, finally, accepted. If an operation fails, a well-defined investigatory procedure follows. A failure can be something as simple as a torn shipping package or something more complex, such as the discovery of a misalignment during an installation operation. Manuals for submarine manufacturing describe the proper procedures for dealing with such failures. The manuals contain clear, step-by-step instructions that an engineer can follow to decide what actions to take next.

Most of the time the investigatory procedures are sufficiently robust to rectify a failure. Sometimes, however, the investigatory procedures also fail. In the submarine industry, this is called a nonconformance. We can think of a nonconformance as an atypical problem—a problem for which the usual investigatory procedures do not return a solution. An example of a nonconformance is when a correction of a misalignment problem results in another misalignment problem.

When a nonconformance occurs, the cognizant engineer confers with colleagues and superiors until a solution is found. The solution is documented and filed in a paper system in the department that handled the nonconformance. Unfortunately, there is not a public repository for nonconformances that engineers with similar problems in other departments can exploit. Consequently, many nonconformances are researched and resolved in duplicate.

The obvious solution to the problem is to collect all experiences with nonconformances in one place and provide a means to retrieve similar, resolved nonconformances when a new one occurs. The first approach taken by General Dynamics was to represent these experiences in a rule-based ES. Existing paper documents of nonconformances were collected and translated directly into rules. For example, each document consisted of 35 slots—item, manufacturer, location,…,problem, solution. The rules were simple translations of a document where the if part of the rule consisted of the first 34 slots and the then part was the solution.

Although the initial results were promising, the ES was shown to be inadequate before it reached deployment in the plant. The system would fail if a new nonconformance did not have an exact counterpart in the rule base. Furthermore, attempts to construct generalized rules over sets of similar nonconformances proved to be frustrating.

An internal research project at General Dynamics suggested that a CBR approach might be a better match for the task. The insight that motivated the CBR approach was that existing nonconformance documents simply looked more like a case than a rule. If one could map the existing documents directly into a case-like structure, then perhaps one could construct retrieval and adaptation algorithms to do the rest.

The suggestion resulted in a CBR application named Cebrum that was deployed during the second quarter of 1990 in three of six engineering departments that are responsible for submarine fluid systems. Up until the time of deployment, it was estimated that $750,000 had been spent on finding a way to alleviate the redundant engineering problem. This figure included work on the analysis of the problem, work on the initial ES solution to the problem, the research project, and the development of the CBR system. The return during the first year of using Cebrum was estimated to be $240,000.

4.3.2 The Structure of Cebrum

The structure of Cebrum is shown in Figure 4.8. It is a straightforward CBR system consisting of five modules.

The Input Module

This module consists of a case-acquisition submodule that is used to build and edit the case library and a problem-acquisition submodule that is used to submit outstanding nonconformances to the system. The structure of a case models previous paper reports of nonconformances and their solutions.

The Reasoning Module

This module consists of retrieve and adapt submodules. The retrieval of previous similar nonconformances is guided by a set of relevance rules. A relevance rule correlates a particular kind of nonconformance with a subset of the slots in a case. An example is this:

   If the problem is incorrect pressure then look at item, type, manufacturer, and fluid type.

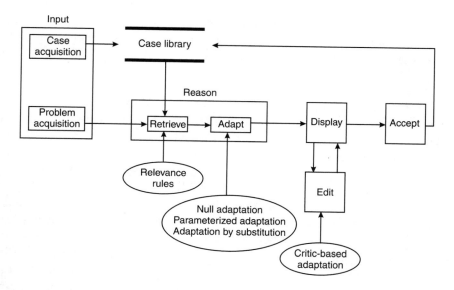

Figure 4.8 The structure of Cebrum.

The kinds of adaptation techniques used are null adaptation, parameterized adaptation, and adaptation by substitution. An example of parameterized adaptation is this. Consider a source case whose problem is incorrect valve pressure. If there are slots in the source case that correspond to an actual pressure X, a desired pressure Y, a number of quarter turns Z needed to achieve the desired pressure, and an equation in the solution slot calculating Z as a function of X and Y, then a similar valve with the same problem will use the equation in the source case to propose a solution to the new problem.

A simple form of adaptation by substitution occurs when multiple cases in the library are applicable to a new problem. The field over which adaptation occurs is the solution field. For example, sometimes the features of a new problem impose constraints on admissible solutions. Suppose the system were able to propose three solutions—to repair a valve, to replace a valve, and to consult the manufacturer of a valve—in that order. If the user had indicated that the valve is nonrepairable in the problem description, the second solution will be substituted for the first one.

The Display Module

This module displays the case that describes the new problem, the potential solutions found by the reasoning module, and the relevance rule for each case that was used to retrieve the solution. The user then chooses to accept or edit one of the solutions offered.

The Edit Module

This module is the manual counterpart of the reasoning module, where adaptation is performed by the user (or critic) instead of the reasoning module. In addition, the edit module is the primary means to maintain the system. Maintenance usually involves fine-tuning a relevance rule if it is found to be faulty.

To illustrate, consider the following example. For a problem of incorrect valve pressure, the system recommended a solution of repairing the valve. In the source case where the solution was found, the slot for the manufacturer of the valve was left blank. However, the user knew that this particular kind of valve was nonrepairable and had to be replaced. The user changed the solution in the new case to *replace valve* and made sure that the name of the manufacturer was included in the case. Then the slot manufacturer was added to the body of the relevance rule that was used to retrieve the case to make sure that the manufacturer's name would be considered for similar problems in the future.

Thus, the next time a problem with incorrect pressure is submitted to the system, the case with the solution *replace valve* is retrieved if there is a match with the name of the manufacturer, everything else being equal. Otherwise, the case with the solution *repair valve* is retrieved.

The Accept Module

After everything is said and done, the user can accept the work or abort the exercise. If acceptance is chosen, the module prints a standard non-conformance document and enters the new case and any revised relevance rules or adaptation techniques into the case library.

4.3.3   Features: The Maintenance of the System

Software development generally involves a strain between system developers and users. Accordingly, it is important to establish user requirements and expectations early in development.

This strain is compounded when one is designing problem-solving methods in a system that in some way mimics the problem-solving activities of the user. The reason is that human problem solving is a rather private affair and is often hard to articulate. What usually happens is that the developer to some extent becomes the kind of problem solver for which the system is being developed, or else the user becomes the kind of problem solver that is implemented in the system. The easier it is for one of these two things to happen, the more likely it is that the system will be successful.

During the development of Cebrum, the latter happened. Since the CBR interpretation of the task was a close match to current practice, users had little difficulty adapting to the system and learning how to maintain the system themselves. By design, this happened because early in the development of the system, the users were coached on the CBR framework for problem solving and shown the design options for a CBR application. The key to these discussions was to focus on simplicity and understandability. If a technique seemed to confuse users or to be unconvincing, the technique was discarded although it might have been the right way to go.

4.3.4　Implications for Network Management

The experience with Cebrum echoes a good, familiar lesson: CBR is not a panacea for a complete problem-solving task and can often be embedded as a component of a larger problem-solving system. For example, the first line of attack on failures in submarine manufacturing operations is left untouched. Engineers still consult the usual investigatory procedures in their manuals when a failure occurs. These procedures are more or less written in stone as a result of years of experience. The CBR system is invoked only if these procedures are inconclusive.

An additional lesson that can be gleaned from the application is to focus on simplicity and understandability when selecting the techniques to use in a CBR system. It is easy and tempting to become overambitious when designing a CBR application. However, a task achieved with simple methods has a better chance of success. In addition, as demonstrated in the development of Cebrum, it is likely that the maintenance costs can be reduced considerably when users of the system can maintain the system themselves.

Finally, it is always a good idea to compare the existing method of solving a problem with available techniques, including rule-based ESs, CBR systems, and other methods. The application described here owes part of its success to the observation that a nonconformance form on paper has the appearance of a case rather than a rule. With this simple insight, everything else fell into place naturally. In contrast, if the task had been to automate the process of finding solutions to problems by following instructions in manuals, the better approach would have been to use the ES paradigm of problem solving. This approach is sometimes called electrifying manuals.

4.4　SMART: A CBR CALL SUPPORT SYSTEM

4.4.1　Background

Compaq Computer Corporation manufactures personal computers. The company's increase in business in recent years called for the establishment of a Customer Support Center to handle customer requests and problems. The center was opened in March 1991. Its charter was to resolve customer problems and requests over the phone.

The operations required to process a call—from the initial receipt of the call to closing the call—are shown in Figure 4.9. There are two paths through the process. The path 1–2–3 corresponds to updating the customer about the status of a problem. The path 1–2–3–4–5 corresponds to finding a solution to a problem and relaying the solution back to the customer. The goals are to decrease the number of traversals of the first path and to increase the timeliness of traversals of the second path. In plainer words, the goal is to solve customer problems over the phone when customers first call so that they will not have to call back.

The usual method of performing the operation in step 4 was to research the problem by reading manuals, duplicating the problem, and conferring with fellow employees. Often solutions are simply carried around in support engineers' heads. In a small operation, these methods are feasible.

As the number and variety of problems increase, however, the task of step 4 becomes unmanageable. For example, Compaq's line of products has expanded over the years and ranges from laptop computer systems to high-end systems. In addition, many of Compaq's products are networked in larger, heterogeneous networks—thus introducing a new class of problems. The number of support calls has increased proportionally. Since Compaq's support center opened, the number of support requests more than doubled, and the telephone support group grew by 100%.

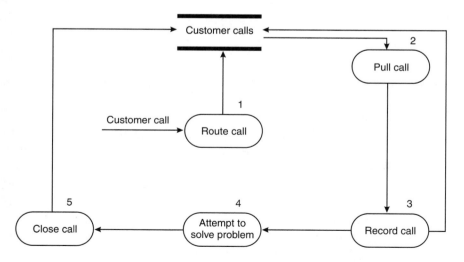

Figure 4.9 The call-handling process.

To cope with the increase in business and the volume of support requests, Compaq attached an optional CBR problem-solving component to step 4, leaving the rest of the system intact. The CBR component, named Smart, was developed with the help of Inference Corporation. An internal study showed that the call-handling procedures without Smart resolved almost 50% of test scenarios, compared with 87% when conjoined with Smart. The system was put into operation by the end of 1991 and paid for itself within a year of deployment.

4.4.2 The Structure of Smart

The structure of Compaq's call-handling system with Smart is shown in Figure 4.10. The subpath 4–8–9 corresponds to the process of retrieving solutions to problems from the case base. The subpath 4–6–7 corresponds to the process of updating the case base.

When a call is received, the support engineer enters basic customer information such as name and address and a freestyle description of the problem into a call-log form on a computer screen. If the solution to the problem is not apparent, the engineer has the option of invoking Smart while the customer is still on the phone. If Smart is invoked, the problem description is copied to a secondary screen and a search is performed over the case base. The search is based on key-term matching, where key terms in the problem description are matched against key terms of the cases in the case base.

The general look of the secondary screen is shown in Figure 4.11. The initial search displays abstracts of a set of ballpark cases that match the current problem. A number from 0 to 100 is attached to each case to indicate its degree of relevance given the information provided so far. The search also displays a set of questions whose answers would further prune the initial set of cases. The engineer then asks the customer for answers to these questions. As the answers are entered, the search algorithm prunes the original list of matching cases and adjusts the cases' degrees of relevance. If a degree of 70 is reached for any retrieved case, the engineer relays the solution to the customer. Otherwise, the problem is declared an unresolved problem, and the phone session with the customer is terminated.

The unresolved problem scenario triggers off-line research into the problem by senior engineers. When a solution to the problem is confirmed, the problem is developed into a case and entered into the case library for future reference (see path 4–6–7 in Figure 4.10).

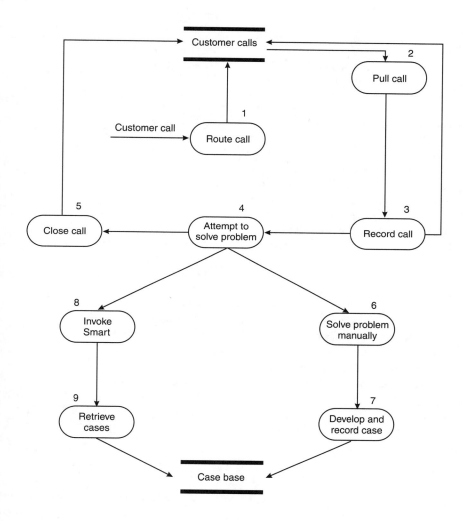

Figure 4.10 The call-handling process with Smart.

4.4.3 Features: A Case Construction Style Guide

The engineers who develop the case follow a Case Construction Style Guide. On the one hand, the guide provides the engineers with suggestions on how to phrase questions and when to use list, text, yes/no, single/multiple, and numeric answers. In addition, the guide explains good practices for asking questions—for example, it advises engineers to ask the most general questions first.

Description	Get Description	
Experiencing spurious problems in a Compaq file server resulting in a lockup situation, Ethernet topology, high traffic contention.		
Questions about This Problem	Browse Question	
How do you categorize the problem (pick the first that applies)? What network operation system are you using? What major release of Banyan Vines are you using? Which family of Compaq machines is it? Are you using Vines SMP? Are there any interrupt conflicts? Does the server lock up and then have to be turned off?		
Matching Cases:	Browse Case	Show Actions
[35] Banyan, Vines 4.10 Systempro server and PCs locking up. [35] 99 Netware 386/286, client hangs, token ring, IBM LAN support. [34] Banyan and Systempro w/ multiple server panics and locks up #1. [33] 97 Netware 386, lost or spurious IRQ, 48 processor. [33] Banyan and Systempro w/ multiple server panics and locks up #2. [32] 98 Netware 286, server locks on boot, 486 processor.		
Search Case Base New Search	End Search	Unresolved Search

Figure 4.11 A Smart screen.

The guide also helps to ensure consistency in the case base by laying out the terminology to use when describing problems and solutions and the terminology to use when formulating questions and possible answers. These instructions are crucial because the key-term matching algorithm depends upon consistency in the language.

4.4.4 Implications for Network Management

Of the four applications discussed in this chapter, Smart is most applicable to problem solving in network operations. The case base, for example, is partitioned with respect to the following domains: Novell, LAN Manager, Banyan, Unix, DOS, Windows, OS/2, and general hardware and software.

When one is designing a case base for a network-management operation, it is likewise good practice to partition the case base with respect to domains of interest in the network. The domains of interest will certainly vary over network enterprises—so there is not one correct partition.

A good rule of thumb is to examine the distribution of responsibilities in the existing network enterprise and model the design of the case base after this. For example, in some network enterprises the responsibilities are distributed with respect to location—Tom is responsible for the L.A. operation, and Jane is responsible for the Chicago operation. In other domains, responsibilities are distributed with respect to subnets or with respect to operational tasks—Tom is responsible for the QA network, and Jane is responsible for the corporate network.

An interesting phenomenon in network operations is the following: Natural experts emerge, become known, are identified with their area of expertise, are appealed to over and over again by their peers, and are typically dragged away from their assigned duties. Tom is our expert on PC problems, Jane is the expert on problems with Solaris operating systems, Sam is the expert on Unix scripts, and so forth. This phenomenon is, of course, natural and occurs in all human enterprises.

It follows that a reasonable way to partition a case base is with respect to natural pockets of expertise that have emerged in the network operation. If possible, the natural experts in each partition could be assigned the additional tasks of researching the unresolved problems and then developing a well-structured case that represents the problem. Over time, the expertise will solidify in the case base, and, thus, experts will be dragged away from their assigned duties less often.

4.5 SUMMARY

This chapter has described four deployed CBR applications. The chapter demonstrates some of the ways in which CBR methods have been used in fielded operations. Each application was described in terms of the task to be achieved, the payoff of implementing the task with CBR methods, how the methods were integrated into existing operations, and system maintenance. We also provided discussions of the CBR methods' implications for network management.

The purpose of the chapter is to get secondhand experience that we can carry to the network-management domain. Some of the important lessons that we should remember as we proceed to the next chapter are these:

1. CBR is not appropriate for all problem-solving tasks.
2. CBR can be embedded as a small component of a larger problem-solving system.
3. It is not necessary to use each operation in the CBR paradigm.
4. One should focus on simplicity and understandability when designing a CBR system.
5. It is important to design a domain language for a CBR application up front.
6. One should always keep in mind the costs of maintaining a CBR system.

At this juncture, we will pay special attention to the task of managing faults in networks. We will examine a tool that is good at detecting and localizing network alarms (Section 5.2) and another tool that is good at managing the flow of work required to oversee the repair of the faults and service requests (Section 5.4). Both of these are off-the-shelf tools—the network management platform Spectrum and the network TTS Action Request System. Finally, we will show how one can integrate these tools and enhance their combined capabilities with CBR methods. The result will be a fairly complete solution to the task of network fault management.

4.6 FURTHER READING

For further details on Prism see Goodman's "Prism: A Case-Based Telex Classifier" in *Innovative Applications of Artificial Intelligence*. The seminal paper on the method of transforming a table into a decision tree is Quinlan's "Learning Efficient Classification Procedures and the Application to Chess End Games" in *Machine Learning*. Quinlan's more recent "Inductive Knowledge Acquisition: A Case Study" in *Applications of Expert Systems* presents a thorough discussion of the method. Also see Rissland and Skalak's "Combining Case-Based and Rule-Based Reasoning: A Heuristic Approach" in *Proceedings of the International Joint Conference on Artificial Intelligence* and Barletta's "An Introduction to Case-

Based Reasoning" for further discussions of using these algorithms in CBR systems. Goodman's "Automated Knowledge Acquisition From Network Management Databases" in *Integrated Network Management II* is a good example of using the algorithm to derive rules from network management databases.

For further details on Canasta, see Register and Rewari's "CANASTA: The Crash Analysis Troubleshooting Assistant" in *Innovative Applications of Artificial Intelligence*. For further details on Cebrum, see Brown and Lewis's "A Case-Based Reasoning Solution to the Problem of Redundant Resolutions of Nonconformances in Large-Scale Manufacturing" in *Innovative Applications of Artificial Intelligence*. For further details on Smart, see Acorn and Walden's "SMART: Support Management Automated Reasoning Technology for Compaq Customer Service" in *Innovative Applications of Artificial Intelligence 4*.

Other good examples of deployed CBR applications are provided in Kolodner's book, *Case-Based Reasoning*, which contains an appendix of about 80 abstracts of CBR systems, including both research and fielded systems.

A good way to find out about other CBR applications is to perform a computer search on CBR at a university library or to "surf the World Wide Web."

Select Bibliography

Acorn, T., and S. Walden, "SMART: Support Management Automated Reasoning Technology for Compaq Customer Service," in *Innovative Applications of Artificial Intelligence 4*, Menlo Park, CA: AAAI Press, 1992.

Barletta, R., "An Introduction to Case-Based Reasoning," *AI Expert*, August 1991.

Brown, S., and L. Lewis, "A Case-Based Reasoning Solution to the Problem of Redundant Resolutions of Nonconformances in Large-Scale Manufacturing," in *Innovative Applications of Artificial Intelligence 3*, Menlo Park, CA: AAAI Press, 1991.

Goodman, M., "Prism: A Case-Based Telex Classifier," in *Innovative Applications of Artificial Intelligence 2*, Menlo Park, CA: AAAI Press, 1991.

Goodman, R. M., "Automated Knowledge Acquisition From Network Management Databases," in *Integrated Network Management II*, I. Krishnan and W. Zimmer (eds.), Amsterdam: North Holland/Elsevier Science Publishers, 1991.

Kolodner, J., *Case-Based Reasoning*, San Mateo, CA: Morgan Kaufmann, 1993.

Quinlan, J., "Learning Efficient Classification Procedures and the Application to Chess End Games," in *Machine Learning*, R. Michalski, J. Carbonell, and T. Mitchell (eds.), Palo Alto, CA: Tioga, 1983.

Quinlan, J., "Inductive Knowledge Acquisition: A Case Study," in *Applications of Expert Systems*, J. Quinlan (ed.), Turing Inst. Press, 1987.

Register, M., and A. Rewari, "CANASTA: The Crash Analysis Troubleshooting Assistant," in *Innovative Applications of Artificial Intelligence 3*, Menlo Park, CA: AAAI Press, 1991.

Rissland, E., and D. Skalak, "Combining Case-Based and Rule-Based Reasoning: A Heuristic Approach," in *Proceedings of the International Joint Conference on Artificial Intelligence*, Los Altos, CA: Morgan Kaufmann, 1989.

Part III
Network Management and
Case-Based Reasoning

At this juncture, we should have a good understanding of the CBR framework for problem solving. In addition, we should have a feel for what makes a CBR application successful. We are now ready to carry these ideas over to the task of fault management in the networking domain.

Generally, fault management consists of three broad phases: fault detection, fault diagnostics, and recovery management. In our system, the fault diagnostics phase is supported by CBR. However, fault detection and recovery management are outside the scope of CBR. Thus, our fault management system will include other components in addition to the CBR application.

In Chapter 5, we describe the overall architecture for the fault management system. The system includes an NMP, an automatic trouble ticket generator, a TTS, and CBR. We discuss each of these components in detail, showing how they are integrated to provide a complete, workable solution to the fault management task.

In Chapter 6, we show how to implement the architecture. Our implementation uses Spectrum from Cabletron Systems as the NMP, the Spectrum Alarm Monitor from Cabletron as the automatic trouble ticket generator, and the Action Request System from the Remedy Corporation as the TTS. We demonstrate two approaches by which to incorporate CBR into the system. The first approach implements CBR as a component of the TTS. The second approach implements CBR as a component of the NMP.

Case-Based Reasoning in Network Fault Management Systems

5

In Chapter 5:

❐ *An Architecture for a Network Fault Management System*
❐ *The Network Management Platform*
❐ *The Automatic Trouble Ticket Generator*
❐ *The Trouble Ticket System*
❐ *Critter: A Case-Based Reasoning Trouble Ticket System*

This chapter describes the overall architecture of our network fault management system, the components of the architecture, and the interfaces between each component.

5.1 AN ARCHITECTURE FOR A NETWORK FAULT MANAGEMENT SYSTEM

The fault management architecture consists of five components, of which CBR is a single piece (see Figure 5.1). The main components are an NMP, an automatic trouble ticket generator (ATTG), a TTS, a CBR system layered around the TTS, and a network troubleshooter.

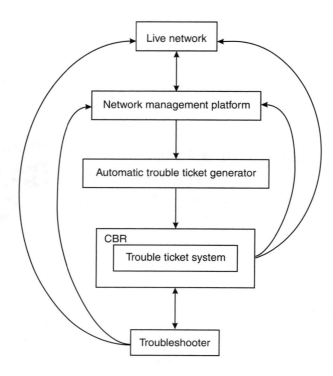

Figure 5.1 The general structure of a CBR fault management system.

The operations of each component are as follows: The NMP detects alarms on the network; the ATTG collects information about the alarms and creates corresponding trouble tickets in the TTS; and the CBR component analyzes the alarms, passing its results to a troubleshooter. The vertical arrows in Figure 5.1 represent the flow of real-time information that results from these operations.

In some instances, the CBR component will be able to execute repairs by sending instructions to the NMP or by performing operations directly on the network, in which case we have a closed-loop fault management system without human intervention. This is shown by the right-most curved arrows in Figure 5.1. More likely, a troubleshooter will execute the repairs offered by the CBR system. This is shown by the left-most curved arrows in Figure 5.1. In both cases, the curved arrows represent historical information that is reapplied to outstanding, real-time alarms.

Two important observations of the architecture are in order: Normally, the TTS is used as a vehicle for managing network service requests as opposed to a vehicle for analyzing a network alarm. Typically, the TTS is used to expedite a service request by passing the trouble ticket through the right channels in the network operation. However, we will enhance the TTS with CBR methods of problem solving that will aid in alarm analysis. At some point, the trouble ticket will be forwarded to an expert repair person who will try to determine a cause of (or at least an explanation for) an alarm. If the alarm was caused by a network fault, the repair person will try to determine ways to repair the fault. These tasks are difficult and time consuming, and it is at this point that the CBR component comes into play.

Second, let us observe that the CBR component is part of a distributed problem-solving architecture: Instead of being the total solution to the fault management task, CBR provides an important piece of the task. The role of the CBR component is to analyze alarms: to determine whether the alarm is caused by network faults and to offer possible repairs for faults. The other components in the system play equally crucial roles, and CBR works in tandem with them.

The instantiation of the architecture that we will describe in this chapter includes Spectrum from Cabletron Systems as the NMP, the Action Request System (ARS) from the Remedy Corporation as the TTS, and the ARS Gateway from Cabletron as the ATTG. The CBR layer is implemented by enhancing the ARS with CBR methods of problem solving. However, let us note that other instantiations are possible. For example, the system that we describe in this chapter includes a TTS that is enhanced with CBR techniques. Alternatively, it would be possible to take a generic off-the-shelf CBR tool and enhance it with TTS techniques.

In Chapter 6, we will work through the development of the fault management system from an implementation perspective. We wish to illustrate that each of the components in the architecture can be replaced with other off-the-shelf tools or *homegrown* tools, provided that they have the appropriate interfaces and hooks to realize the system. The requirements of such interfaces and hooks will be provided during the course of the chapter. Such components that meet the requirements are sometimes called *pluggable* systems.

Let us now proceed to our discussions of the NMP, the ATTG, the TTS, and the CBR component of the architecture.

In Section 5.2, we will examine the functions of the Spectrum NMP and describe how alarms are detected and passed to other applications.

In Section 5.3, we will look at the ATTG, discussing the alarm flooding problem and several ways to alleviate it.

In Section 5.4, we will study the functions of the TTS, using the ARS as an example. In addition, we will explain how CBR methods are incorporated into a TTS.

Finally, in Section 5.5, we will propose several designs for a case-based TTS, ranging from a simple design to a complex design. We will also describe a phased methodology for incorporating CBR into a TTS.

5.2 THE NETWORK MANAGEMENT PLATFORM

In this section, we will see why we use an NMP in a network operation. A sketch of the NMP functions of Spectrum is provided. Although we will be looking at the big picture of an NMP, we will be interested primarily in the particular functions that contribute to our fault management architecture—specifically, how the NMP detects faults on a network and passes the information to a TTS.

5.2.1 Motivation

Some of the common network management tasks include upgrading device operating systems, upgrading the applications running on a device, establishing access permissions for users, or simply determining whether a device is alive. A network manager must also keep an eye on things such as segment bandwidth usage and user complaints.

Clearly, a network administrator could manage a network by physically moving around the network and using the appropriate management tools for each device. However, this approach is impractical.

This is where the concept of an NMP comes into play. Suppose the administrator could manage each network device from a single console. Figure 5.2 shows a high-level network topology as shown in the Spectrum NMP. The administrator clicks on an icon in the topology and brings up options to perform the management tasks of interest.

Thus, the concept of an NMP provides a means of centralized network management or what is sometimes called integrated network management. For large networks that contain hundreds of kinds of devices, including

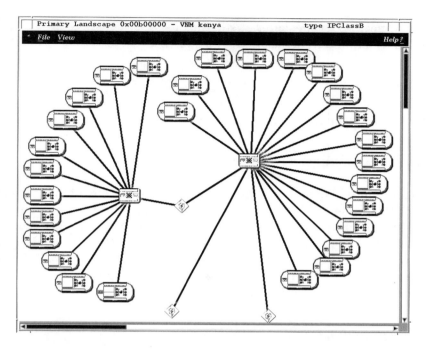

Figure 5.2 A network topology.

routers, bridges, and hubs—each with their own particular management techniques—an NMP is quite useful and perhaps indispensable.

5.2.2 Basic Functions

The concept of an NMP opens the doors for several useful management functions. For example, the Spectrum NMP includes the following functions:

- Show a topological view of the network;
- Show hierarchical views of the network;
- Show logical views of the network;
- Speak and listen to each component of the network;
- Show performance views of the network;
- Configure network components;
- Discover a network;

- Log network events and statistics to a log file;
- Show alarms on the network;
- Speak and listen to other network management applications.

Each of these functions contributes to network fault management. They are described in the following sections.

Show a Topological View of the Network

An important function of an NMP is to depict the structure of the network. Generally, the network is depicted at the component level. In the simplest form, the network is shown as a graph with nodes and arcs. The nodes are labeled by device names such as workstation1, router12, snmp_device_wiz, user_lewis, and so forth. This is the ordinary, commonsense representation of the network. An example of a topological view of a network is shown in Figure 5.2.

Show Hierarchical Views of the Network

In very large networks, the topological view becomes enormous. A common way to make sense of a large number of network components is to form levels of abstraction over the network. Suppose one is looking at the topological view of the network. In Spectrum, a simple *go up* function invokes a view of the same network as a collection of subnets. Further, one can *go down* a level of abstraction. If one wished to go down into a router, one would invoke a view that shows the router's ports and connectivity.

Show Logical Views of the Network

Often it is useful to divide the representation of a network into logical views. A logical view is any partition of the network that helps one understand the structure and dynamics of the network conceptually.

An obvious logical partition is with respect to location. For example, one may desire to examine the portion of the network that is located in floor 3 of building A. Another kind of partition is with respect to network functions. One may wish to examine a view of the portion of the network that is used by the QA lab, the research lab, or the sales organization.

Speak and Listen to Each Component of the Network

The topological, hierarchical, and logical views of a network could be implemented with any sophisticated graphical package. For that matter, one could represent these views with pen and paper. It is a different matter when we want to communicate with the components in the network. This function is perhaps the most basic and important function of an NMP. However, it is difficult to implement—for three reasons:

First, the NMP has to be a part of the network in order to communicate with the rest of the network. This implies that the NMP has to speak the same language as the various network components or else have translation protocols in place that allow the NMP to communicate with devices that speak different languages.

For example, a standard protocol has evolved called the simple network management protocol (SNMP). Any network device that is SNMP-based (or has an SNMP agent) communicates easily with an SNMP-based NMP. Otherwise, the NMP should have an open applications programming interface (API) that allows one to implement a special-purpose protocol for communicating with non-SNMP devices.

Second, the NMP needs to know the kinds of questions to pose to a network component and the kinds of answers expected to be received. Generally, vendors of network devices publish a management information base (MIB) for their device. The MIB can be thought of as a list of questions for which the device provides answers at any point in time. Some of the questions are static (version number and IP address of the device, for example). Other questions are dynamic (for example, parameters that have to do with the health of the device).

Finally, the function of speaking and listening is difficult, as there are really no well-established standard communication protocols or standard MIBs. The most common protocols and MIBs have simply *caught on* in the industry (for good reasons), but there are always competitors that might catch on in a bigger way. For example, competitors of SNMP are SNMPv2 and CMIP. The main competitors of the MIB are MIB-II and RMON. The best an NMP vendor can do is to cover as much of the industry as possible and provide an open API that a vendor or third-party developer uses to implement special-purpose communication protocols. For example, Spectrum provides such an API.

Since fault management ultimately depends upon the ability to communicate with network devices, these issues are important.

Show Performance Views of the Network

This function is related to speaking and listening to network devices. The Spectrum NMP, for example, allows the user to click on an icon device and bring up a performance view of the device. The performance view is a real-time graphical representation of the device's health, where health is a function of the values of MIB variables. This sort of information is quite useful for fault management.

Configure Network Components

This function is also related to speaking and listening to network devices. In this case, however, we want to be able to configure the devices from the NMP console. Some of the ways in which we would configure a device are to download an upgraded operating system or an enhanced MIB or to set watches on some of the MIB parameters. For example, we might want to change the IP address of each device in a segment of a network.

In large networks, these tasks are quite tedious and expensive. If we had 50 routers, each of which needed an upgraded MIB, it would be convenient to do this in one fell swoop from the NMP and to verify that the operation was successful. Similarly, it would be convenient to be able to reassign IP addresses to hundreds of network devices at once. The Spectrum Configuration Manager allows one to perform tasks such as these.

Discover a Network

Suppose we have a large network that has previously been managed piecemeal. Now, however, we want to replace the piecemeal management of the network with an NMP. It would take several days to create representations for each device on the network and several more days to organize the devices into topological, hierarchical, and logical views.

The function of discovering a network performs the task with much greater accuracy and in less time that the manual method. For example, one might discover a device not known to exist, or one might find out that devices thought to be connected are not actually connected. The usual way to do this is to attach the NMP to a router in a network, whereupon the NMP reads the router's network map. If the router is connected to other routers, the NMP automatically attaches to those routers and reads their maps and so on recursively until the network is discovered.

Note that network discovery is useful not only when one wishes to consolidate network management but also when one wishes to track add-ons, moves, and changes in the network configuration.

Log Network Events and Statistics to a Log File

We saw a small portion of a network log file in Figure 4.6. The network log file is like a systems log file for a workstation but on a grander scale. It is a repository for recording all events that occur on the network. Network events include a device connecting to or disconnecting from a network, a threshold having been overstepped, loss of contact with a device, and permission denials.

Although most network events are routine, it is useful to keep a record of them. Note also that an event log typically grows large very quickly and depletes the resources of the device where the log resides. Spectrum solves this problem by archiving the network log file periodically, usually on a daily basis. A month's worth of network events can be unarchived and turned into useful information with Spectrum's reporting facilities using data compression algorithms. These reports may be used to track down and discover network faults.

Show Alarms on the Network

It is useful to make a distinction between network events and network alarms. A network event issues from a network device or an NMP, either by predefined traps or polling. In Spectrum, an alarm corresponds to an event that has been predefined as having an alarm status. Thus, all alarms are events, but not all events are alarms.

For example, a few events that indicate a load threshold being exceeded is cause for concern but not cause for panic. On the other hand, a steadily increasing network load over several days, intermingled with numerous events that indicate network disconnections, should have the status of an alarm. The alarm is probably indicative of a fault and calls for immediate action.

As another example, consider that the loss of contact with a device is not something to worry about. Users frequently unplug their workstations by intention or by accident, and they are plugged back in shortly thereafter. However, the simultaneous loss of contact with all devices on a subnet is more likely to have alarm status and indicate the existence of a fault.

Network administrators can examine an event log with the goal of discovering network alarms and faults; however, this task is tedious and time consuming. The NMP performs a good portion of this function. Since the NMP shows a graphical representation of the network at several layers of abstraction (e.g., as topological, hierarchical, logical, and performance views), it is an easy matter to portray alarms by coloring the icons that represent network components. A common scheme is to color alarms as yellow, orange, or red where red alarms indicate maximum severity.

However, note that two administrators who oversee different network operations would probably have different ideas about which events should have alarm status or the severity that should be assigned to an alarm. One administrator's alarm is often another administrator's nuisance. For this reason, rather than dictating an event or alarm mapping scheme, an NMP should be user-configurable. Accordingly, it is desirable for an NMP to come off-the-shelf with a tried and true scheme of mapping network events to alarms and a mechanism that allows network administrators to customize the NMP for a particular network operation.

The ability of an NMP to show alarms on a network takes us a long way toward the goal of efficient fault management. The missing piece so far is the ability to find the causes of alarms, where causes may include genuine network faults, user-error, or simply FAD. In combination with a TTS, an ATTG, and CBR, we provide the missing pieces. These components are discussed later in the chapter.

Speak/Listen to Other Network Management Applications

An NMP should provide the means to interface with other network management applications in a hierarchical structure (e.g., as a *manager of managers*). The highest level of the hierarchy may concentrate on providing an overall graphical picture of the network with an overview of alarm conditions and performance data. The administrator can then drill down to lower level management applications to see detailed information or to issue device-specific commands.

In addition, the NMP should provide the means to interface with peer applications. The motivation for this feature is the well-known fact that no one vendor provides all the solutions to network management. Vendors are usually specialists in one aspect of network management—for example, in network simulation, network monitors and analyzers, and TTSs. A vendor who tries to cover all aspects of network management is likely spreading itself too thin.

The integration of multiple network management products is analogous to *passing the ball*. One player (i.e., product) carries a task as far as possible. When the task transforms into something that the player cannot handle, the relevant information is passed to another player. Note that we are describing such an approach to fault management, where the players are an NMP, an ATTG, a TTS, and CBR.

One important challenge for this feature is to understand how multiple *players* combine their expertise. The challenge is to design a distributed problem-solving architecture and show how each piece of the architecture contributes to the overall solution. A second, more practical challenge is to provide interfaces that allow the players to exchange data. The latter is a prerequisite for implementing our fault management architecture, assuming that the architecture has been validated. We will show how one meets this goal in Section 5.2.4.

5.2.3 Discovering Network Faults

Discovering a network fault is easy in some instances but quite difficult in other cases. This subject is worthy of a book in its own right. In this section, however, we aim simply to understand the problem and to examine ways in which an NMP helps to alleviate the problem. First, we will make some simple observations.

NMP-Originated Alarms and User-Originated Alarms

Let us note the simple distinction between alarms reported by an NMP and alarms reported by users. The distinction is rather obvious but helpful, nonetheless. NMP alarms show up graphically on the NMP console by coloring or flashing the devices on which the alarms occur. Since such alarms have been predefined, they may be called objective alarms. User-originated alarms correspond to conditions that an individual perceives as being abnormal and, thus, may be called subjective alarms. The vehicles for subjective alarms are typically phone calls to the network administrator or submissions of trouble tickets to a TTS or help desk.

Mapping Alarms to Faults

Now let us get into the mindset of thinking in terms of mapping network alarms to network faults. Anytime we see an alarm, whether it has come from the NMP or from a user, we should ask the question, "Is there a fault?" and then "What is it?" Eventually, of course, we want to think about how to repair the fault. However, prior to considering how to repair it, we should try to determine whether there really is one.

The reason that one should hesitate is that many alarms do not indicate faults. Such alarms are much like temporary knocks and ticks that one might hear in an automobile engine. We may show some concern at first, but then they disappear and we forget that they ever happened. The same holds true for little knocks and ticks that we experience in our bodies or little confrontations that we experience in our dealings with other people. There are no serious faults in these instances. They are just temporary alarms that fortunately go away.

Intermittent Alarms

The phenomenon of temporary knocks occurs in networks also. We call them intermittent alarms. The classic example is an alarm that is caused by a brief surge in network traffic. Another good example is the contact loss alarm that was described in the previous section. A good NMP should be able to filter out these intermittent alarms or, at least, offer a way for a user to tell the NMP not to worry about them.

Bona Fide Alarms

On the other hand, some alarms are bona fide alarms. However, note that the distinction between intermittent alarms and bona fide alarms is relative. Generally, if a contact lost alarm remains in effect for more than a few hours, then it should be considered as possibly indicating a fault that needs attention. If a cash cow application is running on a workstation, then an administrator might be inclined to consider any alarm as indicative of a fault and want to do something about it immediately. This also holds true for critical network devices such as hubs and routers. We would not want either of these devices to be inoperable for more that a few seconds.

In these examples, the mapping between alarms and faults is rather straightforward. The mapping hinges primarily on the length of time that the alarm remains in effect. In other instances, the mapping is considerably more complex and difficult. In the industry, these are called alarm correlation problems.

The Alarm Correlation Problem

Let us first illustrate the alarm correlation problem with an example and then provide a definition.

Suppose that three simultaneous alarms appear on an NMP console as shown in Figure 5.3. The nature of each alarm is described as follows:

Alarm 398 on server *S*—overloaded traffic throughput

Alarm 401 on router *R*—overloaded traffic throughput

Alarm 426 on subnet backbone *B*1—overloaded traffic throughput

In addition, suppose that several alarms have been reported by users on subnet *B*2 saying that their client applications running against S are prohibitively sluggish.

This is not an unlikely scenario in networking operations. The important question is "What is the fault?" The most likely hypothesis is that users have been added to subnet *B*2 that have to access *S* and that the bandwidth of *B*1 cannot accommodate the extra traffic. The solution might be to move all clients on *B*2 that access *S* onto *B*1, to increase the bandwidth of *B*1, or to move *S* to *B*2.

Whatever the best solution might be, note that no one alarm by itself is enough to indicate the fault. The fault is not with *S*, not with *R*, and not

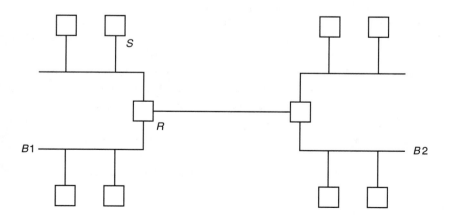

Figure 5.3 A simple alarm correlation problem.

necessarily with $B1$. It would be poor judgment to try to find a repair for each of these alarms in isolation without considering the bigger picture.

These sorts of problems are called alarm correlation problems. The goal is to infer a network fault (or faults) from multiple network alarms (and reports from users). Let us stop to think for a moment about where the burden of alarm correlation should take place.

Ordinarily, the alarm correlation task is performed by expert troubleshooters. However, some types of alarm correlation tasks are performed by the NMP. For example, suppose we have a large set of alarms—one of which occurs on a router and the rest of which occur on all the devices that are on the subnet side of the router. It is not difficult for an NMP to determine that the likely fault is with the router. The Spectrum NMP, for example, would suppress all the alarms except the one on the router.

In the three alarm example in Figure 5.3, however, it would be conceptually and practically difficult for the NMP to determine the fault. Conceptually, the reasoning processes of expert troubleshooters who are good at this sort of problem are not well-understood. Of course, it would be simple to build an if-then rule that would apply to this particular scenario. And, as we know, if the domain were relatively constant and there were no surprises, then this approach would work. But few networking domains are like this.

Practically, if we try to implement the expert troubleshooter's reasoning processes in the NMP, we can easily overload it. The NMP is doing a lot as it is. In plain words, things will get messy if we try to implement too much in the NMP. For this reason, an alternative approach naturally suggests itself: Let us configure the NMP so that it hands off alarm information to a separate application, perhaps residing on a different machine. The external application, then, will carry the burden.

In fact, this is just our approach to the fault management architecture discussed in this chapter, where the separate application is a TTS. The obvious requirement for this approach is that both applications have to have the proper hooks and interfaces in place that allow them to communicate with each other. We will take this up in the next section.

5.2.4   Passing Alarm Information to External Applications

How do we get alarm information out of an NMP and into an external application? Generally, there are two approaches to this task—the command line interface (CLI) approach[1] and the application programming interface (API) approach. See Figure 5.4 in which we use the Spectrum NMP and the ARS as examples of separate applications that we wish to integrate.

The CLI Approach

The CLI approach to passing data from application A to application B requires a set of procedures that allows the operation of B from the command line and a set of hooks in A that allows the execution of these procedures.

For example, the normal way to enter a trouble ticket into the ARS database is to use the user interface (see the ticket form in Figure 5.5). A less elegant way is to enter the ticket from the command line. If the ticket held nine fields, we would be able to submit a ticket by the following command:

```
submitTT f1 f2 f3 f4 f5 f6 f7 f8 f9
```

A hook in Spectrum allows the command to be executed with the appropriate values for the parameters. The command is executed automatically (i.e., unsolicited) whenever an alarm occurs in the network or executed on demand via a button on a device icon in the NMP.

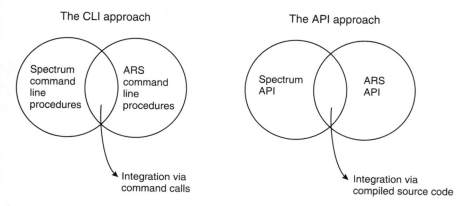

Figure 5.4 The CLI and API approaches to data exchange.

---

1. The CLI method is sometimes referred to as the platform external interface.

The API Approach

The API approach to the exchange of data between two applications requires a programming language (such as C or C++) API for each application. A vendor offers an "open systems" application with libraries, source code, and documentation that a third-party developer uses to build an integrated application. The developers select modules and functions provided in each API as they see fit. With this approach, the applications are integrated at the source-code level.

Of course, it is possible to use both methods in a single integration. The ARS Gateway, for example, uses the API approach for automatic trouble ticket generation and the CLI approach for manual trouble ticket generation.

Trade-Offs and Discussion

The CLI and API approaches to application integration are vehicles that we can use to develop intelligent network management solutions, including our fault management architecture and others. However, there are important trade-offs between the approaches.

The benefits of the CLI approach are:

- Procedures are written in any programming language.
- Procedures are tested and executed in isolation.
- A first-order integration is achieved quickly.
- There is little lost if the project is abandoned.

The drawbacks of the CLI approach are:

- Performance may not be optimal (e.g., response time may be prohibitively slow).
- The integration is limited by the set of procedures and hooks available in each application.
- There could be a problem with security.

The benefits of the API approach are:

- One achieves better performance.
- The integration is not dependent upon available procedures and hooks.
- Security is designed into the integration.

The drawbacks of the API approach are:

- The two applications must have compatible APIs.
- The development time is comparatively long.
- There is a fair amount lost if the project is abandoned.

A viable method of integration that combines the best of both approaches is to use the CLI method to develop a prototype integrated system. If the performance or functionality of the system is prohibitive, then the prototype is used as a specification for building a more efficient integration using the API approach.

In the future, it is likely that the CLI approach to application integration will become increasingly prominent as a means to build proof-of-concept systems. An advantage of the approach is that it allows architects to experiment online with integrated solutions, with less emphasis on implementation details.

With the basic concepts and functions of the Spectrum NMP under our belts, let us proceed to the ATTG piece of the architecture.

5.3 THE AUTOMATIC TROUBLE TICKET GENERATOR

We have laid most of the groundwork that we need to understand the function of an ATTG. The ATTG sends relevant information to a TTS whenever the NMP detects an alarm on the network. What shows up as a colored, flashing device on the NMP console is transformed into a trouble ticket in the TTS. We can implement an ATTG using either the CLI approach or the API approach.

There are a few problems with the idea of an ATTG that, fortunately, can be surmounted rather easily. The main problem is that if a trouble ticket is generated for every network alarm, then the TTS might become flooded with trouble tickets very quickly. Let us call this the alarm-flooding problem.

5.3.1 The Alarm-Flooding Problem

The alarm-flooding problem is caused in part by intermittent alarms. For example, we would not want to see a trouble ticket for every instance in which someone unplugged a workstation.

The problem is also caused by cascading alarms. An example of cascading alarms is when a real alarm instigates a number of apparent alarms, such as when a failed router causes a flood of alarms on devices to which the router is connected.

Furthermore, the problem is caused by flickering alarms. If an alarm occurs every time a threshold of a parameter is overstepped, and it happens that the value of the parameter is oscillating around the threshold, we are flooded with hundreds of identical trouble tickets that differ only with respect to their time stamps.

A possible consequence of the alarm-flooding problem is extra network traffic. If the NMP and the TTS reside on different machines in the network, there could be unnecessary traffic generated whenever alarm information is passed over the network from the one machine to the other. For network alarms such as traffic overload, the use of an ATTG might be a partial cause of the problem.

There are several ways to alleviate the alarm-flooding problem. One approach depends on the extent to which the NMP can perform alarm correlation. Some NMPs are good at this, but others are not. In the failed router example, the Spectrum NMP suppresses the cascaded alarms and shows only the router alarm. Thus, the alarm-flooding problem is less severe.

Another approach is to replace the ATTG with a manual trouble ticket generator. The idea here is to collect all alarms in an alarm buffer. The user then examines the buffer and selects those alarms for which a ticket should be generated and discards the rest. In some network operations, however, this approach defeats the purpose. For example, if we want to know immediately when a router goes down we would not want to spend extra time sifting through an alarm buffer.

A similar approach deals with the problem on the TTS side. All alarms are forwarded to an initial buffer in the TTS. In the TTS industry, these are called incident reports, as opposed to bona fide trouble reports. The user selects the incident reports that should be escalated to trouble tickets proper and discards the rest. Note, however, that this solution does not solve the problem of extra network traffic.

The best solution to the alarm-flooding problem is to enhance the ATTG with a filtering mechanism. This approach solves the problem in one fell swoop and provides a network administrator with the ability to set ATTG policies. This is the approach taken by Cabletron's ARS Gateway, the ATTG that integrates the Spectrum NMP and the ARS TTS.

5.3.2 ATTG Filters

The idea of alarm filtering is straightforward. We wish to control the alarms that are transformed into trouble tickets. Several filtering schemes are possible, but we will describe just one below.

We will allow filtering to take place over the following seven alarm parameters:

- Device;
- Device type;
- IP address;
- Location;
- Alarm severity;
- Alarm type;
- Alarm age.

The meaning of these parameters should be obvious, save perhaps alarm age.

First, recall that (a) an event issues from a network device, either by predefined traps or polling; and (b) an alarm corresponds to an event that has been predefined as having an alarm status. In Spectrum, such alarms show themselves as colored, flashing icons.

Now, other events may trigger an action to remove the alarm. A simple example is that of unplugging a workstation, resulting in a contact lost alarm (say a red alarm) flashing on the Spectrum icon that represents the workstation. When the workstation is replugged, an event occurs that triggers an action to remove the alarm, causing the icon to return to nonalarm status (i.e., green). Alarm age is the amount of time that elapses from the assertion of an alarm to the removal of the alarm.

The alarm age filter takes on a numeric value from 0 to 1440 (24 hours) that shows the minutes that an alarm is allowed to age before it is transformed into a trouble ticket. The alarm age filter essentially tells an alarm, "You have to be X minutes old before I will recognize you as a bona fide alarm and generate a trouble ticket." Or in different words, "If X minutes have elapsed since the event that changed the status of the icon to abnormal, and no event since then has changed the status of the icon back to normal, then generate a trouble ticket."

The alarm age parameter is very useful for filtering out intermittent alarms and flickering alarms. For example, one may want a trouble ticket to be generated for an unplugged workstation only if the workstation remains unplugged for more than say five hours.

We will let alarm severity take on the values yellow, orange, and red to indicate increasing severity.

Finally, we will allow each parameter to be preceded by a tilde "~", which means *except*; and we will reserve a keyword all that serves as a value for each parameter (except alarm age). The meaning of all is obvious—it is a "wild card."

A filter is a concatenated list of parameter values of the following structure:

```
device : device type : IP address : location : alarm severity
: alarm type : alarm age
```

A few examples will help to clarify this scheme. Consider the following:

```
all : all : all : all : all : all : 0
```

(let all alarms generate trouble tickets immediately)

```
all : all : all : all : all : all : 5
```

(let all alarms generate trouble tickets if they are not removed in five minutes)

```
all : Cisco_rtr : all : all : all : all : 0
```

(let all alarms for Cisco routers (only) generate trouble tickets immediately)

```
all : ~WS_SGI : all : all : red : all : 0
```

(let only red alarms for all devices except SGI workstations generate trouble tickets immediately)

We define the behavior of the ATTG by putting together a list of filters. For example:

```
all      :WS_SGI    :all    :all        :red :all :15

all      :WS_IBM    :all    :all        :red :all :15

all      :WS_DEC    :all    :all        :red :all :15

~doc     :WS_DEC    :all    :all        :all :all :0

all      :all       :all    :~Bldg2     :all :all :0

all      :CiscoRtr  :all    :all        :all :all :0
```

This means that we want red alarms for all SGI, IBM, and DEC workstations to generate trouble tickets if they are not removed in 15 minutes (except for doc, which should never generate any tickets). Do not generate any tickets that occur in Building 2, but do generate tickets for any alarms on Cisco routers immediately.

There are a few minor problems with this particular scheme. Fortunately, they are worked out rather easily. Suppose an administrator devises a set of filters that is inconsistent. For example:

```
all      :WS_SGI :all    :all    :red    :all                  :15

coffee :WS_SGI :all    :all    :red    :all                  :0

all      :all       :all    :all    :all    :~contact lost :0
```

If the alarm contact lost occurs on coffee, we have two kinds of inconsistencies at play. The first two filters are inconsistent with respect to alarm age. The first one says to generate a ticket if the alarm is not removed in 15 minutes and the second one says to generate it immediately. A second inconsistency occurs because the first two filters say to generate a ticket conditionally with respect to alarm age, but the third filter says not to generate a ticket for any contact lost alarm.

Several strategies are used to resolve conflicts among competing filters. The simplest perhaps is the strategy, "Smaller alarm ages always override larger ones." Given this strategy, a trouble ticket would be generated immediately in the example above. This is the strategy used in the ARS Gateway.

An alternative strategy is the following:

1. Smaller alarm ages always override larger ones.
2. A do not always overrides a do.
3. If rule 1 and 2 conflict, 2 overrides 1.

Given this strategy, a ticket would not be generated in our example. The rationale for the strategy is the commonsense notion that the exception always overrides the rule.

5.3.3 ATTG Policies

By extending the alarm filtering function, we provide another useful service for fault management. A particular set of filters is considered a policy that can be put into effect at different points in time. Thus, an administrator defines an 8:00 a.m.–5:00 p.m. policy, a 5:00 p.m.–8:00 a.m. policy, a weekend policy, and vacation policies.

This feature is especially useful when considered in conjunction with alarm notification. By alarm notification, we mean the medium by which a network troubleshooter is notified about a network problem. Thus far, we have assumed that alarm information is forwarded to a TTS. However, it is quite possible to forward the information to other applications such as e-mail and paging systems.

We build upon a filter by adding two parameters that indicate the notification medium and whom to contact. For example:

```
all  :WS_SGI     :all   :all   :red    :all :15 :TTS      :sam

all  :WS_SGI     :all   :all   :red    :all :15 :e-mail :tom

all  :CiscoRtr   :all   :all   :all    :all :0  :page    :jane
```

An e-mail notification might be appropriate for filters in an 8:00 a.m.–5:00 p.m. policy when the troubleshooter is likely to be at a workstation. A paging notification might be preferred for filters in nighttime, weekend, and vacation policies.

This additional feature requires separate e-mail or paging applications. In addition, the applications would need the proper hooks and procedures so that they could be integrated with Spectrum, as was described in Section 5.2.4.

5.4 THE TROUBLE TICKET SYSTEM

We have looked at two pieces of our fault management architecture thus far—the NMP and the ATTG. Let us now examine the TTS component of the architecture.

Recall that fault management consists of three phases: alarm detection, alarm diagnostics, and fault recovery. Alarm detection is performed by network users and the NMP. Typically, alarm diagnostics and fault recovery are performed by network troubleshooters.

Recently TTSs have been introduced to assist with the fault recovery phase. Once an alarm is detected, and a repairable fault is discovered and prioritized with respect to other faults, sooner or later it will come time to recover from it. At this juncture, an administrator should be aware of the criticality of the fault and the steps required to fix it and know who should be charged with the task, know at any time the status of the repair operation, and be able to verify that a repair has been made. This is where the TTS is useful—to impose a structure on the recovery process and to provide a means to keep tabs on progress.

In this section, we will discuss the motivation and functions of a TTS, using the Action Request System TTS as an example. After this, we will be ready to describe the final piece of the fault management architecture—CBR. The CBR piece will complete the picture by assisting with the fault diagnostics phase.

5.4.1 Motivation

Why use a TTS in fault management? The best answer is that without structure and policy it is easy for somebody to drop the ball during the recovery process. Under the worst scenario, there are several individuals required to execute different pieces of a repair and one of the individuals, for whatever reason, does not complete his or her task. This slows down and confuses the whole process.

However, if we have a well-structured work-flow process in place and a means to keep tabs on overall progress, we increase the probability of a timely and successful repair. This is where the TTS comes in to play.

5.4.2 Basic Functions

The functions of the TTS are as follows:

- Represent the information needed to expedite repairs in a trouble ticket;
- Provide the means to prioritize, schedule, and dispatch work tasks;
- Provide the means to oversee the repair process;
- Provide means of communication among repair participants;
- Clocking;
- Statistical analyses;
- Reporting;
- Provide long-term memory.

Each of these functions contributes to fault management. However, we will be interested most in the first and last of these functions—the representation of information and long-term memory. If the procedures and results of the repair process are represented adequately and placed in long-term memory, then we are set for extending the TTS with CBR.

Represent Information Needed to Expedite Repairs in a Trouble Ticket

There are a number of core fields that are present in a trouble ticket:

- Ticket ID;
- Time and date of problem detection;
- An abstract of the problem;
- Severity of the problem;
- Name of the person opening the ticket;
- Ticket status.

The choice of additional fields will vary with respect to the characteristics of a particular network operation. Other candidate fields include:

- The subnet(s) or user(s) affected;
- The addresses of the machines and network devices involved;
- Alarm ID;
- Alarm time and date;
- A record of ticket updates and modifications;
- Resolution of the problem;
- Repair priority;
- Schedule of tasks;
- Repair person;
- Notification method.

For example, Figure 5.5 shows the structure of a Spectrum trouble ticket in the ARS.

It is important to note that there are generally two approaches to the choice of fields to include in a ticket.

The first approach is the incomplete but customizable ticket. With this approach, it is assumed that there cannot be a universal ticket that is applicable to all network operations. A ticket contains a few core fields

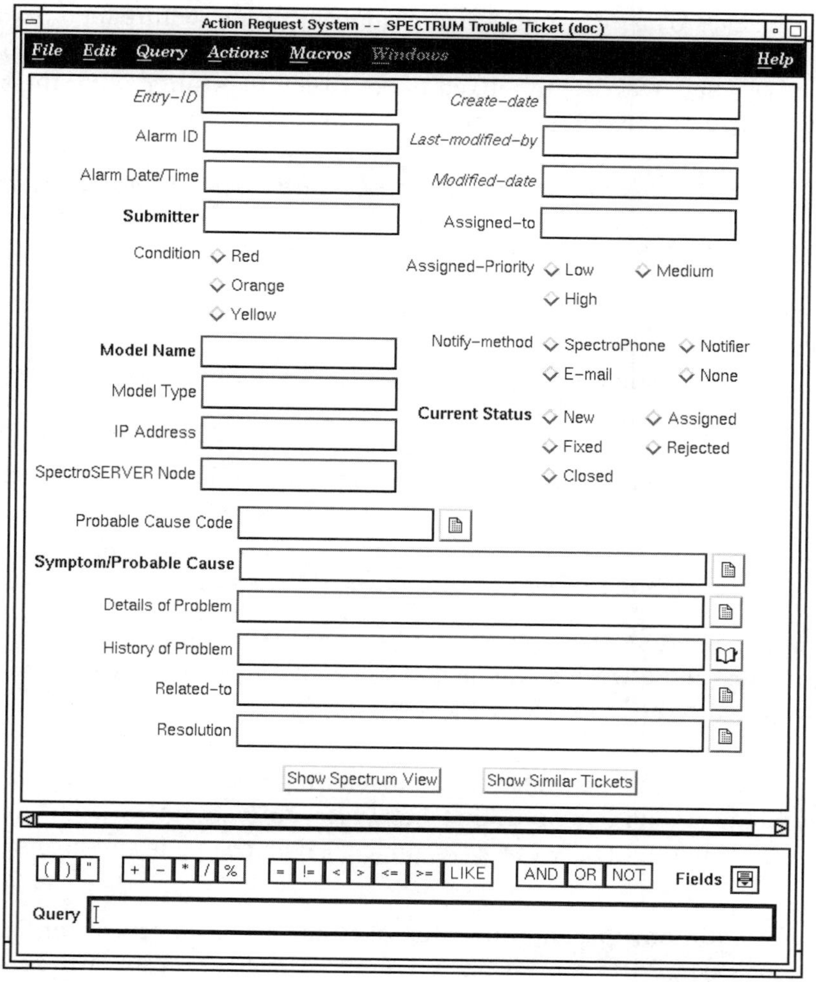

Figure 5.5 An example of a trouble ticket.

but provides the means to create any number of other fields, including labels, data types, positioning in the ticket, and the means to control the visibility of fields for people who will be looking at the ticket.

The second approach is the complete but uncustomizable ticket. The assumption here is that experience has shown what is essential in a ticket. Taking this approach, the network administrator exploits the experience and avoids the extra expenses of TTS customization.

The trade-offs between the two approaches are clear. If an administrator wishes to analyze and design a trouble ticket structure for a particular network, the former approach is better. On the other hand, if one is convinced that a TTS with a fixed ticket structure is suitable for the task, then the latter approach is better.

Provide the Means to Prioritize, Schedule, and Dispatch Work Tasks

This function hinges around the following fields:

- Repair priority;
- Schedule of tasks;
- Notification method.

Using these fields, the administrator is able to construct plans for repairing faults on the network and to control the order in which the plans are carried out.

Provide the Means to Oversee the Repair Process

With this function, an administrator has the means to perform spot-checks on progress in network maintenance. For example, one can query the TTS database with the following sorts of requests:

- Show me the schedules of all outstanding tickets with a priority of 1.
- Show me the schedules that are at least 75% complete.
- Show me all of Sam's task assignments, both complete and incomplete.

In addition, the administrator has the means to revise the schedule—for example, to reprioritize the faults waiting for a fix or to reassign tasks.

Provide Means of Communication Among Repair Participants

Since a trouble ticket is the main repository for all activities regarding a repair, the TTS alleviates communication problems among those who are involved in the repair. For example, if the only means of communication were the phone, we can imagine the possible communication breakdowns when there are an administrator and several troubleshooters, users, and checkers who play a part in the repair. However, if each participant were instructed to communicate via the trouble ticket, these breakdowns would be less likely.

The ARS can be configured to send messages to other participants whenever certain actions occur on the ticket. One places a demon on certain fields so that a change in value triggers a notification to another participant. Consider the field *ticket status* and its legal values:

> ticket status: open
> accepted
> rejected
> diagnosed
> assigned
> in progress
> resolved
> verified
> closed

A demon on ticket status is constructed so that whenever its value equals *assigned*, a notification is forwarded automatically to the assignee. Also, an administrator can construct a demon such that whenever the value of ticket status changes, the person in charge of overseeing the repair process is notified.

TTSs that allow one to construct such demons belong to the incomplete but customizable category. This is the main method by which an administrator imposes structure and policy on the repair process.

Clocking

One sense of clocking is to measure the time that elapses with respect to several dimensions of the repair process—for example, from report to diagnostics, from diagnostics to start of the repair, and from start of the repair to completion. This is usually achieved by looking at the time-stamps each time the ticket is modified.

Another sense of clocking is a sort of reminding. If one has hundreds of repairs in the queue, it is easy for some of them to fall through the cracks. With clocking, however, one instructs the TTS to remind one if, for example, the ticket has not been touched in 24 hours.

Statistical Analyses

The former sense of clocking allows the administrator to gather statistics on repairs. Some of the statistics that are useful are as follows:

- What is the average time required to fix the fault "error in IP address?"
- What is the average time for Sam to do it?
- What is the average time for Jane to do it?

Other useful statistics might involve a single network device. One might wish to know the faults that have occurred on a certain vendor's router since the time that the router was put into operation.

Reporting

Given that we can tap in to several aspects of the state of network maintenance and gather statistics, it is useful to be able to make hard copies of the data. The ARS, for example, allows a user to design the hard copy at the TTS console and send it to a printer. An alternative approach is to allow one to export the data out of the TTS and then process the data in a separate, high-end reporting package.

Provide Long-Term Memory

After all is said and done regarding a repair, it is useful to retain this information for future use. For example, long-term memory is required to gather statistics and to provide reports on maintenance histories. In addition, as should be obvious by now, the solutions to past problems are potentially useful for similar future problems. This feature is crucial for the CBR piece of our architecture, and we will visit it again in Section 5.5.

5.4.3 Trouble Ticket Systems as CBR Problem Solvers

In this section we want to advance to thinking of a TTS as a potential CBR problem solver. Given our discussion of the general CBR framework in Chapter 3 and our discussion of TTSs above, this should not be difficult. For example, it is easy to conceive of a trouble ticket as a case and a TTS database as a case library. Let us try now to determine the additional information and functions that are needed in order to transform a TTS into a genuine CBR problem solver.

Most of the information we need already exists in a trouble ticket. Recall that cases exhibit three primary components:

1. A primary slot that is the main question.
2. A number of slots that represent background information.
3. A method or formula by which to determine an answer to the main question, given the background information.

In a network trouble ticket the primary field (#1) is typically an abstract of the problem. We interpret this field as the question, "This is the problem; what is the diagnosis?" or, "This is the problem; what is the prescription for fixing it?" The resolution field is reserved for the answer to the question.

The objective of the CBR component is to appeal to such past resolutions in the case library when proposing a resolution for a new problem, in much the same way that an expert troubleshooter appeals to prior experience. Let us look more closely at how an expert troubleshooter deals with a trouble ticket and compare his actions to the CBR framework of problem solving.

All the fields in the ticket are fair game for diagnosing the problem. These fields are considered background information (#2). The troubleshooter, however, will not be interested in all of them. The fields used for management purposes probably will not help much in finding a solution to the problem.

Furthermore, some of the fields will give the troubleshooter ideas about where to look for more information, other questions to ask, and tests to perform. That is, the fields in the ticket serve as a springboard for further inquiry. The extent to which a troubleshooter advances beyond the information in a ticket is what makes him an expert troubleshooter. Unfortunately, this is just the information that is usually left out of a trouble ticket, although it is important for future problem solving.

Component #3 in a CBR problem solver comprises this latter kind of information. The troubleshooter somehow manages to take a mental leap from the abstract of the problem and other information in the ticket to a resolution. Such mental leaps are hard to articulate, but if we find some way to express this in a trouble ticket, we will have achieved a great deal.

Finally, CBR requires retrieval and adaptation methods. When a new problem is recorded in a ticket and assigned to a troubleshooter, it would be useful to be able to retrieve resolutions to similar problems and adapt the old solutions (if necessary) to fit the new problem. We should note that retrieval and adaptation contribute to #3. Part of the mental leap of the troubleshooter involves appealing to past experience and applying this experience to a new problem, and this is precisely the objective of CBR.

In sum, we should concentrate on the following points when we consider the transformation of a TTS into a CBR problem solver:

1. A well-defined language by which to express problems and solutions;
2. Component #3—a representation of the means by which a solution is found;
3. Retrieval methods;
4. Adaptation methods.

We will have a better chance of implementing these functions in a TTS if we have one that belongs to the incomplete, but customizable category. Nonetheless, we still may have a chance at realizing a simple CBR system in complete, but uncustomizable TTSs. In some instances, all we need is a retrieval method to invoke the ways in which similar problems have been solved in the past.

5.4.4 On Intelligent Agents

In this section, let us digress briefly to discuss the philosophy of building intelligent agents and to show where our approach to network fault management fits into the picture. We will draw upon research and accomplishments in the AI and robotics communities.

The work on implementing intelligence in a machine since the early 1970s is an interesting story. We will not tell all of it here. Instead, we will discuss an important turn in the field that occurred in recent years.

In the 1970s, it was thought that intelligence could be partitioned into different pieces—for example, into perception, learning, planning, acting, and natural language understanding. Different groups of researchers focused on just one piece of intelligence in hopes that if their piece could be solved, and other groups solved their pieces, then all that was left would be to put the pieces together.

This approach is called the classical architecture for realizing an intelligent agent. It is well-known that the approach did not work. The general feeling now is that the different pieces of intelligence are inextricably bound and interleaved. It is hard to conceive of any one piece of intelligence in isolation from the rest of the pieces. Furthermore, it is difficult to demarcate the different pieces of intelligence in the first place.

An alternative approach to understanding intelligence, called the subsumption architecture, was conceived in the mid 1980s. The idea is that intelligence consists of layers of functionality, where each additional layer subsumes the layers beneath it and demonstrates increasingly complex behavior. An important insight of this new approach is that higher layers of functionality do not interfere with the layers beneath it. Thus, if a higher layer becomes dysfunctional for whatever reason, the lower layers would remain operable.

The classical example in the robotics community is the task of obstacle avoidance. Consider a layer of functionality of order 0: traveling. Given a starting point A and a destination B, the robot is able to calculate a path from A to B and proceed to B. Next, consider an order 1 behavior stopping, darting, and traveling. If the robot perceives that it is not moving (probably because it hit an obstacle), it darts away randomly, recalculates, and continues traveling to point B. Now consider an order 2 behavior sensing, darting, and traveling. The idea is that whenever the robot senses a foreign object in close proximity, it darts away from it, recalculates, and proceeds to point B.

The order 2 behavior depends upon the order 1 behavior, which, in turn, depends upon the order 0 behavior. However, if the order 2 behavior becomes dysfunctional—for example, if the sensors were turned off—it is still possible that the robot could make it to B with the order 1 behavior.

It is useful to think of TTSs enhanced with CBR in this way. We have described the functions of TTSs and have indicated how we will build on some of these functions in order to transform the TTS into a CBR problem solver. It is important to note that we do not want the CBR layer to interfere with the existing functionality of the TTS. We would be defeating our purpose if we allowed this to happen.

With this simple insight, let us get back on track and show how to enhance a TTS with CBR.

5.5   CRITTER: A CBR TROUBLE TICKET SYSTEM

Our pet name for the idea of a CBR trouble ticket system is Critter. The idea itself connotes the transformation of a TTS into a CBR problem solver. However, there are several ways to do this, some of which are more complex than others. Below we discuss the various design options that are available to us, discuss a simple design and a complex design, and propose a methodology for building a Critter-based system.

5.5.1   An Array of Designs

In Chapter 3, we discussed each component of a CBR system—the concept of a case, various retrieval and adaptation methods, and possible ways in which to structure a case library. Figure 3.11 summed up the options for implementing each component in a single picture. Figure 5.6 shows these options carried over to the TTS domain.

For purposes of discussion, the main components of the system are labeled with numbers, and the options for implementing each component are labeled with small letters. These labels will be used to refer to any specific design. For example, the system 1a/2ae/3a/4d refers to the system that uses key-term matching as the retrieval method, null adaptation and critic-based adaptation as the adaptation methods, manual execution as a means to carry out solutions, and master cases as the structure of the case library.

Clearly, there is quite a large number of possible designs that one could build into a TTS. The question now is how to select a design for a particular application. The answer is partly constrained by the character of the TTS.

Consider the structure of a TTS database. Practically all commercial TTSs store their tickets sequentially. This is not a prohibitive constraint, but it might increase the time it takes to search the case library looking for similar tickets. Very few TTS have mechanisms that group tickets with respect to some specific tickets field, for example, problem type.

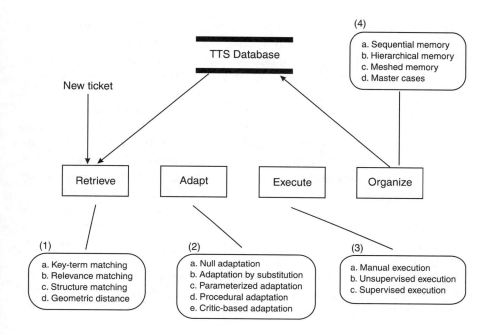

Figure 5.6 Design options for Critter.

Moreover, about half of the TTS packages have a fixed ticket structure. This imposes limitations on the language that is used to talk about the network domain. Nonetheless, most tickets have at least a representation of *problem* and *solution* that we might be able to exploit with CBR techniques.

Finally, less than half of the TTS packages provide the user with hooks by which to customize the system. Ideally, we would like a TTS that allows the user to place buttons in the ticket that would execute external, user-defined processes. For example, a *retrieve* button would execute a process that looks at the contents of an unresolved input ticket and then displays a list of similar tickets that have successful solutions. An *adapt* button would look at the solution in a retrieved ticket and adapt it for the original input ticket. An *execute* button would execute the solution (if possible) and an *embed* button would enter the newly resolved trouble ticket into the TTS database.

These constraints will not apply if one is building a Critter-like system from scratch or if one's approach is to transform a generic CBR system into a network TTS. Then, however, we would probably have to give up many of the other TTS functions or else reinvent the wheel.

Next, the question of which design options to choose depends on the kinds of information and the resolution methods commonly used in one's operation. If a troubleshooter's style of problem solving makes use of procedures in manuals, then it is a good idea to use procedural adaptation. The troubleshooter develops a master case for each problem type and attaches a general procedure for finding a solution to the problem. If the procedures are implemented in a programming language such as UNIX scripts, one can execute the solution in manual, supervised, or unsupervised mode. Such a design is described by the system 1a/2d/3abc/4d.

In addition, note that one can write procedures that are capable of variable-binding and thus open the possibility of using adaptation by substitution and parameterized adaptation. This design is described by the system 1a/2bcd/3abc/4d.

Below we describe two of the more common designs—a simple, albeit useful, design and a more complex design. Both of these designs can be implemented in rather short order.

5.5.2 A Simple Design

The simplest design is the system 1a/2ae/3a/4a (i.e., a design that uses key term matching, null adaptation and critic-based adaptation, manual execution, and a sequential memory). In Chapter 6, we show how one can implement this design in a Spectrum/ARS integration. Note that this design might be the only option available if the TTS belongs to the complete but uncustomizable category.

The idea is to look at the problem type in a new ticket and retrieve a set of resolved tickets that match on the same problem type. The solutions in the resolved tickets serve as reminders as to how the problem has been resolved in the past. Also, the resolved tickets will show who solved the problem, and, thus, this person can be used as a resource. If none of the old solutions apply, a new solution will be found and entered into the case library for future reference.

An assumption of this design is that one has formulated a set of initial, broadbrush problem types that occur on the network. When one opens a ticket, one has to choose from the set of problem types or else define a new problem type. Since the design depends on key-term matching, it is not advisable to allow freeform English to describe the problems. However, it is admissible to use freeform English to describe the solution.

A second assumption of this design is that the TTS has a mechanism by which to create and display the legitimate values in the problem type field. This is usually done with cascading text.

5.5.3 A Complex Design

In this section, we describe a variety of ways of improving our simple design.

The system described above retrieves a set of ballpark solutions for a particular kind of problem. However, consider the problem type *contact lost*. One would believe that the kinds of solutions for loss of contact with a printer would be different from the kinds of solutions for loss of contact with a server. Furthermore, one would suppose that the kinds of solutions for loss of contact with a server depend upon the type of access to the server—telnet, NFS mount, or ftp. However, in the design above one would be shown the solutions for every instance of the problem *contact lost*.

We fine-tune the set of solutions by using relevance matching. We match over additional fields in order to prune the set of solutions. For example, a single relevance rule as follows would suffice:

If problem="contact lost," then match on "device type" and "access mode"

Furthermore, we find other potential solutions by relaxing some of the constraints in the rule or by adding more constraints. It is reasonable to think that if no solutions were forthcoming from the rule above, there might be other solutions in tickets that match on access mode and location. If none were forthcoming here, there might be solutions in tickets that match on access mode only. This line of reasoning is couched in the following relevance rule:

If Problem Type=contact lost then match on "device type" and "access mode,"
Else stop or match on "access mode" and "location"
Else stop or match on "access mode"

This rule allows one to navigate the case base in search of a solution. The rule imposes a virtual hierarchical or meshed memory structure on the case base, even though the tickets are arranged sequentially. We move around the memory structure by switching the constraints in the rule.

The extra burden of relevance matching is that one has to have the knowledge of where viable solutions might be forthcoming for each problem type and the means to represent the knowledge in a relevance rule. Moreover, one has to have the means to modify the relevance rules as the system evolves.

Now let us consider how to adapt a proposed solution. Recall that in the simple design 1a/2ae/3a/4a, we used the methods of null adaptation and critic-based adaptation. These methods are straightforward. The former method maps solutions directly to a new problem and the latter method allows the troubleshooter to review potential solutions and adapt them to the current problem. Note that critic-based adaptation is a catchall term for any way that the troubleshooter adapts a solution, including adaptation by substitution, parameterized adaptation, and procedural adaptation.

Before we approach more complex adaptation tasks, let us emphasize three important truths:

1. No one adaptation method will be universally applicable to all retrieved solutions.
2. Some ways in which a troubleshooter adapts a solution are inexplicable.
3. A structured language for expressing solutions is imperative.

In addition, let us make sure that we understand the concepts of names, variables, and unification. For our purposes, a name picks out any individual item. "doc" is the name of my workstation. A variable is a placeholder for a name. We generally establish a convention for distinguishing a name and a variable. For example, anything with a "?" in front of it is a variable, otherwise it is a name. Thus, "doc" is a name but "?x" is a variable. Unification occurs when we substitute a name for a variable (a.k.a., variable binding). For example, we can imagine all the unifications for which the relation "?x is in building ?y" is true.

Now, the best approach to the automation of the adaptation task is to examine each problem type one at a time. For each problem type, we look at the solution(s) that have repaired the problem. Next, for each problem/solution pair, we try to find any names that are common to both of them and consistently replace these names with variables. We may call this step "case generalization." Finally, we ask the question "Do any solution variables depend upon variables in the Problem Type field or any other fields in the ticket; and if so, can I create a formula or algorithm that relates the solution variables to the problem variables?"

Let's consider an example of adaptation by substitution to illustrate this. Consider the following two cases, where each case exhibits the same problem type but different solutions (assume that case generalization has taken place):

Problem Type: Workstation ?x cannot copy files to workstation ?y
Solution: Workstation ?x copies files to the /tmp directory on workstation ?y

Problem Type: Workstation ?x cannot copy files to workstation ?y
Solution: Put "?x +" in the /rhosts file on workstation ?y

It is easy to see how new solutions are derived by unifying variables in the problem type field and passing the bindings as solutions.

We would have to repeat this exercise for each problem/solution pair that our CBR system knows about. In some cases, the solution field will contain variables that are included in a formula or in an algorithm. These kinds of solutions correspond to parameterized adaptation and procedural adaptation, respectively.

Furthermore, in our example, the variables of interest were contained in the problem type field; however, it is possible that variables will be contained in other fields also. For example:

Machine: ?x
Problem Type: cannot copy files to workstation ?y
Solution: ?x copies files to the /tmp directory on workstation ?y

Machine: ?x
Problem Type: cannot copy files to workstation ?y
Solution: Put "?x +" in the /rhosts file on workstation ?y

Finally, it is advisable to have a well-structured language in order to make use of case-adaptation methods. We have to worry about little things such as "cannot copy files to" and "can not copy files to." These expressions will not match. Generally, it is more practical to make users accommodate a language rather than to make a language accommodate users. The best way to do this is to use cascading menus.

5.5.4 A Methodology

In this section, we provide a design methodology for incorporating CBR into an existing TTS package. Our guiding principles are:

1. Start with a simple design.
2. Design new functionality in increments.

It is all too easy to be overambitious too soon. However, if we adhere to these principles, we will avoid getting into trouble.

The methodology has been partitioned into three phases with respect to increasing complexity and challenge:

I   Simple Retrieval
II   Complex Retrieval
III   Adaptation

It is not always necessary to follow through each phase. For many network operations, Phase I could prove to be very useful and quite inexpensive, and one could stop there.

Before we describe the methodology, let us make the following notes:

- Each phase of the methodology will require modifications to the TTS. Before implementing a design, it is necessary to verify that the TTS has the means to make the modifications, either by using an API or special tools provided with the TTS. We will discuss implementation details in Chapter 6, using Spectrum and the ARS as examples.

- Each phase builds upon earlier phases. However, neither phase interferes with earlier phases, nor does it interfere with the functions of the TTS. Recall that this is the motivation for the concept of a subsumption architecture (see Section 5.4.4).

- Most CBR systems start with a few cases that hold broadbrush representations of problems and solutions in their domain. The goal is to have the case base become increasingly refined, with expanding domain coverage, as it is used. If such an initial case base is not available, one will have to create one. (We should note that, unlike expert systems, the case base does not have to be correct in beginning.) In the methodology below, we assume that an initial case base is not available.

- Finally, a word about who actually builds and oversees the CBR system is in order. Should this person be the system administrator, a troubleshooter, a user (or users) of the system, or a CBR specialist/developer? Our goal is to build a system that does not require a CBR specialist or extensive maintenance (e.g., see the discussions of other applications in Chapter 4). Phase I, for example, is simple enough to be initiated by users of the system. The more challenging methods in Phases II and III, however, will require experienced troubleshooters to develop the cases.

Phase I—Simple Retrieval

1. Make a list of common problem types and implement the list as a cascading menu in a TTS. This will force a user to be consistent with the language for expressing network problems. However, one should allow a user to enter a new problem type if the problem is not already in the list. In addition, it is a good idea to accompany the problem field with an Other Info field that contains free-form English. For example:

   > Problem Type: contact_lost_with_server
   > Other Info: the server is indy2 and my machine is doc. this problem has been happening intermittently for several days.
   > Solution:

2. Implement the simple version 1a/2ae/3a/4a. The user will be able to retrieve all prior tickets that have recorded a solution to the problem type in the outstanding ticket. Some TTS packages allow the user to place a button in the ticket and to define the functionality of the button (e.g., the ARS). The button keys off the value in the Problem Type field in order to collect a set of related tickets. Essentially, the button says, "Show me all tickets that exhibit this kind of problem and have solutions."

Phase II—Complex Retrieval

1. Make a list of relevance rules. To do this, look at each problem type and ask the question, "What additional information would help me determine a solution to this problem?" Put the informa-

tion in a table with two columns. Each row of the table represents a problem type and a list of slot names. Table 5.1 provides a good example:

Table 5.1
Complex Retrieval

Problem Type	Relevant Information
contact_lost_with_server	client_machine, server_machine, access_mode, subnet
machine_down	machine_name, machine_type, operating_system
cannot_print	machine_name, printer_name, file_type

The first entry in the table is a relevance rule that says:

If the problem is contact_lost_with_server
then look at client_machine, server_machine, access_mode, subnet

2. Add the fields in each relevance rule to the trouble ticket structure. Thus, all of the fields in the Relevant Information column will be included as fields in the trouble ticket. The consequence of this step is that we will have all the information in the trouble ticket that we will ever need, although we will never need all the information for any particular problem.

   Unfortunately, the trouble ticket might become quite large and messy if we did this. In addition, a user might find it difficult to identify the subset of fields for which one should provide values when submitting a ticket.

   The best way around this problem is to associate a pop-up menu with each problem type. The pop-up menu will hold slots for the relevant information and (optionally) a list of legitimate values from which the user selects options.

3. Enhance the retrieval button so that it keys off relevance rules. Define the behavior of the retrieval button so that it searches for tickets that match with respect to the problem type and the relevant information associated with the problem type. We will not

need a perfect match with each bit of relevant information. In that case, we would probably not retrieve any solutions. Thus, we define the retrieval button so that it displays retrieved tickets with partial matches. The worst case is that we retrieve tickets that match only with respect to problem type, which is the same result achieved in Phase I.

Phase III—Adaptation

1. Make a list of relevance rules and their corresponding solutions. To do this, we extend the table in Phase II with an additional column for the solution. The solutions will be expressed in free-form, so we devise a language to express solution types in the same way in which we devised a language to express problem types. We pick out names in the solution that also occur in the other two columns. For example, consider the following:

Problem Type: cannot_print
Other Info: machine_name=doc, printer_name=nlpc,
file_type=postscript, your_server=brat
Solution: you need to copy the printcap file on brat to your machine. then try printing again.

The case developer transforms this ticket into the following:

Problem Type: cannot_print
Other Info: machine_name=doc, printer_name=nlpc,
file_type=postscript, your_server=brat
Solution: copy brat:/etc/printcap onto doc:/etc/printcap

2. If possible, use adaptation by substitution in order to relate a relevance rule to a solution. One way to do this is to use the method of transforming names into variables and unifying the variables in the retrieved ticket with the names in the new ticket. The case developer does this to the ticket above as follows:

Problem Type: cannot_print
Other Info: machine_name=?doc, printer_name=nlpc,
file_type=postscript, your_server=?brat
Solution: copy ?brat:/etc/printcap onto ?doc:/etc/printcap

Now, if we get a new ticket with a similar problem, but where machine_name=gist and your_server=blue, then the following solution is proposed via adaptation by substitution:

Problem Type: cannot_print
Other Info: machine_name=gist, printer_name=nlpc,
file_type=postscript, your_server=blue
Solution: copy blue:/etc/printcap onto gist:/etc/printcap

3. If possible, create a formula or algorithm in order to relate a relevance rule to a solution. To do this, we transform names into variables as before, but we pass the variables to a formula or algorithm for additional computation to arrive at a solution.

For example, let us consider a hard problem, "saturated_host." One way to solve the problem is to look at the other hosts with which the host communicates, measure the average amount of traffic between the host and the other hosts, and make sure that the host is located on the same subnet as the hosts with which it communicates most. Generally, the kind of algorithm that does this belongs to the class of clustering algorithms. Here we will not spell out the algorithm in detail.

If such an algorithm were implemented (call it cluster), then the following case takes a problem "saturated_host" and provides a recommendation:

Problem Type: saturated_host
Other Info: target_host=?doc
subnet=?subnet
connected_hosts={(?x1,?s1,?tx1), (?x2,?s2,?tx2),...,
(?xn,?sn,?txn)}
Solution: advice=cluster(?doc, ?subnet, {(?x1,?s1,?tx1),
(?x2,?s2,?tx2),....,(?xn,?sn,?txn)})

We assume that "cluster" is passed a target host, the subnet on which the target host is located, a list of connected hosts, their locations, and their average amounts of traffic with respect to the target host. The output of "cluster" is a recommendation to move the target host to an alternative subnet.

5.6 SUMMARY

This chapter has described an architecture for network fault management. The goal of the architecture is to expedite the detection and repair of problems that occur on a network. The architecture consists of an NMP, an ATTG, a TTS, and a CBR component built into the TTS. We provided descriptions of each component, using as examples Spectrum from Cabletron Systems as the NMP, the Action Request System from the Remedy Corporation as the TTS, and the ARS Gateway from Cabletron as the ATTG. In addition, we discussed several designs for incorporating CBR into a TTS, ranging from a very simple design to increasingly complex designs. In Chapter 6, we show how to implement the architecture.

5.7 FURTHER READING

The origin of the work presented in this chapter is Lewis's "A Case-Based Reasoning Approach to the Resolution of Faults in Communications Networks" in *Integrated Network Management III*.

For further reading on network management in general and fault management in particular see Carter and Dia's "Evaluating Network Management Systems: Criteria and Observations" and Disabato's "Key Technologies for Integrated Network Management" in *Integrated Network Management III*; Leinwand and Fang's *Network Management: A Practical Perspective*; Olesen's "Network Management in Large Networks" and Westcott's "A Simple Model for Integrated Network Management" in *Information Network and Data Communications II*; and *IEEE Communications Magazine's* "Special Issue: OSI Network Management Systems."

For further reading on network management platforms in general, see Mahler's "Multivendor Network Management—The Realities" in *Integrated Network Management III*. For more detail on the Spectrum network management platform, see Abec, Leischner, and Segner's "Applying Inductive Modeling Technology to Tackle the Problem of Integrated Network Management" in *Integrated Network Management III*.

To learn more about the alarm correlation problem refer to Bouloulas, Cato, and Finkel's "Alarm Correlation and Fault Identification in Communication Networks;" Jackson and Weissman's "Alarm Correlation for Telecommunications Network Surveillance and Fault Management;" and Jordaan and

Paterok's "Event Correlation in Heterogeneous Networks Using the OSI Management Framework" in *Integrated Network Management III.*

Book references for artificial intelligence and network management are Ball's *Network Management With Smart Systems*; Ericson, Ericson, and Minoli's *Expert Systems Applications to Integrated Network Management*; and Liebowitz's *Expert System Applications to Telecommunications.* Also see Goyal's "Artificial Intelligence in Support of Distributed Network Management" in *Network and Distributed Systems Management* and Muralidhar's "Knowledge-Based Network Management" in *Telecommunications and Network Management Into the 21st Century.*

Further readings that focus on the concept of intelligent agents, including the classical architecture and the subsumption architecture, are Maes's *Designing Autonomous Agents: Theory and Practice From Biology to Engineering* and Back and Minsky's *The Society of Mind.* Lewis's "AI and Intelligent Networks in the 1990s and Into the 21st Century" in *Worldwide Intelligent Systems: Approaches to Telecommunications and Network Management* compares intelligent agents in network management with ongoing work in robotics and distributed problem solving.

The best source for learning about TTS design issues is *RFC 1297 (NOC Internal Integrated Trouble Ticket System Functional Specification Wishlist),* published by the IETF User Connectivity Problems Working Group. RFC 1297 is accessible by e-mailing to noc-tt-req@merit.edu.

Related works on adding a CBR component to a TTS are Kriegsman and Barletta's "Building a Case-Based Help Desk Application" and Dreo and Valta's "Using Master Tickets as a Storage for Problem Solving Expertise" in *Integrated Network Management IV.*

Lewis and Dreo's "Extending Trouble Ticket Systems to Fault Diagnostics" discusses enhancements to TTSs in addition to those described in this chapter.

For work on solving the saturated host problem with clustering algorithms (and other methods) see Songerwala's "Efficient Solutions to the Network Division Problem."

Select Bibliography

Abec, M., M. Leischner, and P. Segner, "Applying Inductive Modeling Technology to Tackle the Problem of Integrated Network Management," in *Integrated Network Management III (C-12),* H.-G. Hergering and Y. Yemini (eds.), Amsterdam: North-Holland/Elsevier Science Publishers, 1993.

Ball, L., *Network Management With Smart Systems,* New York: McGraw-Hill, Inc., 1994.

Bouloulas, A., S. Cato, and A. Finkel, "Alarm Correlation and Fault Identification in Communication Networks," Research Report 17967, IBM Research Division, T. J. Watson Research Center, Yorktown Heights, New York, 1992.

Carter, E., and J. Dia, "Evaluating Network Management Systems: Criteria and Observations," in *Integrated Network Management III*, I. Krishnan and W. Zimmer (eds.), Amsterdam: North-Holland/Elsevier Science Publishers, 1991.

Disabato, M., "Key Technologies for Integrated Network Management," in *Integrated Network Management III (C-12)*, H-G. Hergering and Y. Yemini (eds.), Amsterdam: North-Holland/Elsevier Science Publishers, 1993.

Dreo, G., and R. Valta, "Using Master Tickets as a Storage for Problem Solving Expertise," in *Integrated Network Management IV*, W. Zimmer and D. Zuckerman (eds.), Amsterdam: North Holland/Elsevier Science Publishers, 1995.

Ericson, E., L. Ericson, and D. Minoli (eds.), *Expert Systems Applications to Integrated Network Management*, Norwood, MA: Artech House, 1989.

Goyal, S., "Artificial Intelligence in Support of Distributed Network Management," in *Network and Distributed Systems Management*, M. Sloman (ed.), Wokingham, England: Addison-Wesley Publishing Company, 1995.

Jackson, G., and M. Weissman, "Alarm Correlation for Telecommunications Network Surveillance and Fault Management," in *Integrated Network Management III (C-12)*, H.-G. Hergering and Y. Yemini (eds.), Amsterdam: North-Holland/Elsevier Science Publishers, 1993.

Jordaan, J., and M. Paterok, "Event Correlation in Heterogeneous Networks Using the OSI Management Framework," in *Integrated Network Management III (C-12)*, H-G. Hergering and Y. Yemini (eds.), Amsterdam: North-Holland/Elsevier Science Publishers, 1993.

Kriegsman, M., and R. Barletta, "Building a Case-Based Help Desk Application," *IEEE Expert*, December 1993.

Leinwand, A., and K. Fang, *Network Management: A Practical Perspective*, Reading, MA: Addison-Wesley Publishing Company, 1993.

Lewis, L., "A Case-Based Reasoning Approach to the Resolution of Faults in Communications Networks," in *Integrated Network Management III (C-12)*, H.-G. Hergering and Y. Yemini (eds.), Amsterdam: North-Holland/Elsevier Science Publishers, 1993; a version also published in Proceedings of *IEEE INFOCOM '93*, IEEE Computer Society Press.

Lewis, L., "AI and Intelligent Networks in the 1990s and into the 21st Century," in *Worldwide Intelligent Systems: Approaches to Telecommunications and Network Management*, J. Liebowitz and D. Prerau (eds.), Amsterdam: IOS Press, 1995.

Lewis, L., and G. Dreo, "Extending Trouble Ticket Systems to Fault Diagnostics," *IEEE Network*, Vol. 7, No. 6, 1993.

Liebowitz, J. (ed.), *Expert System Applications to Telecommunications*, New York: John Wiley and Sons, 1988.

Muralidhar, K., "Knowledge-Based Network Management," in *Telecommunications and Network Management Into the 21st Century*, S. Aidarous and T. Plevyak (eds.), New York: IEEE Press, 1994.

Maes, P. (ed.), *Designing Autonomous Agents: Theory and Practice from Biology to Engineering and Back*, Cambridge, MA: MIT Press, 1991.

Mahler, D., "Multivendor Network Management—The Realities," in *Integrated Network Management III (C-12)*, H.-G. Hergering and Y. Yemini (eds.), Amsterdam: North-Holland/Elsevier Science Publishers, 1993.

Minsky, M., *The Society of Mind*, New York: Simon and Schuster, Inc., 1985.

Olesen, K., "Network Management in Large Networks," in *Information Network and Data Communications II*, D. Khakhar and V. Iversen (eds.), Amsterdam: North-Holland/Elsevier Science Publishers, 1988.

Pyle, R. (ed.), "Special Issue: OSI Network Management Systems," *IEEE Communications Magazine*, Vol. 31, No. 4, 1993.

Sloman, M., and K. Twidle, "Domains: A Framework for Structuring Management Policy," in *Network and Distributed Systems Management*, M. Sloman (ed.), Wokingham, England: Addison-Wesley Publishing Company, 1994 (especially Part IV and Chapter 18).

Songerwala, M., "Efficient Solutions to the Network Division Problem," *TR-12-94*, Department of Computer Science, Curtin University of Technology, Perth, Western Australia, 1994.

Westcott, J., "A Simple Model for Integrated Network Management," in *Information Network and Data Communications II*, D. Khakhar and V. Iversen (eds.), Amsterdam: North-Holland/Elsevier Science Publishers, 1988.

Wies, R., "Policies in Network and Systems Management—Formal Definition and Architecture," *Journal of Network and Systems Management*, Vol. 2, New York: Plenum Press, March 1994.

Implementing Case-Based Reasoning in Network Fault Management Systems

6

In Chapter 6:

❏ *Implementing CRITTER in a Fault Management System*
❏ *A Generic Cased-Based Reasoning Tool*

In Chapter 5, we defined an architecture for a network fault management system. The components of the architecture are:

- An NMP;
- An ATTG;
- A TTS;
- A CBR module.

The NMP monitors the network, collects network events, and maps and correlates events into alarms. The ATTG forwards alarm information to a TTS in the form of a trouble ticket. The TTS enforces a predefined structure on the activities involved in the management of the alarm, including the activities of troubleshooters, testers, and overseers. An important task in alarm management is to determine whether the alarm indicates a fault and, if so, to determine a repair for the fault. The CBR module automates this step by proposing repairs for alarms based on past experience.

In this chapter, we will instantiate the architecture. The NMP and ATTG components of the architecture are, respectively, Spectrum and the Spectrum Alarm Monitor developed by Cabletron Systems. However, we will describe two modes of implementing the remaining components of the architecture.

In the first mode the TTS component is the ARS developed by the Remedy Corporation. The CBR module is implemented using low-level functions provided in the ARS and is, thus, dependent upon the ARS.

In the second mode, the CBR module is integrated with the NMP and is TTS-independent. This latter mode allows alternative TTSs and other network management applications to tap the problem-solving experience in the CBR module.

In Section 6.1, we provide a tutorial for implementing the fault management system with a case-based TTS. Cabletron Systems has implemented such a system in a product called the ARS Gateway. In addition to CBR, the ARS Gateway integrates other important alarm management functions in Spectrum and the ARS, comprising a complete instantiation of the fault management architecture described in Chapter 5. In the tutorial, we explain and reconstruct the initial steps needed to implement the CBR module.

In Section 6.2, we describe a design for standalone CBR module. The design shows (a) how CBR is integrated into the alarm management facilities of Spectrum, and (b) how other network management applications can exploit the CBR module.

6.1 IMPLEMENTING CRITTER IN A FAULT MANAGEMENT SYSTEM

This section is a hands-on tutorial for implementing the fault management architecture in which the CBR module is implemented in the TTS (i.e., a Critter system).

The tutorial takes about four hours to complete. Ideally, the reader would have Spectrum and the ARS running on a UNIX workstation as we work through the tutorial. However, this is not a requirement. Without these products, it is still useful to work through the tutorial and imagine what is happening by looking at the figures.

It is assumed that at least a portion of a live network is modeled in Spectrum. In addition, it is assumed that the reader has a basic knowledge of UNIX shell programming, the operation of Spectrum and the ARS, and an understanding of the fault management architecture described in Chapter 5.

6.1.1 Spectrum

First, let us start Spectrum and open the network alarms view (see Figure 6.1). In the bottom portion of the screen we see a list of alarms that are currently outstanding. The top portion of the screen is dedicated to a particular alarm, which the user can select by highlighting an alarm in the alarm list.

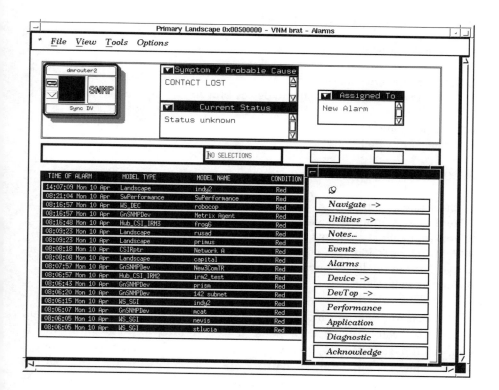

Figure 6.1 A network alarm view in Spectrum.

On the left of the top portion, we see the icon that represents the device on which the alarm occurred and toward the middle of the top portion, we see the description of the alarm. In Spectrum terminology, this is called the "symptom/probable cause" of the alarm. A symptom/probable cause of an alarm is a piece of hard-coded ACSII text that is associated with each type of alarm. This text can be edited by designated users and, thus, can include other hard-coded information (e.g., recommended actions.)

If you were to explode the symptom/probable cause window for this particular alarm, you would see the following:

CONTACT LOST

SYMPTOMS:

Device has stopped responding to polls

PROBABLE CAUSES:

1. Device Hardware Failure
2. Cable between this and upstream device broken
3. Power Failure
4. Incorrect Network Address
5. Device Firmware Failure

RECOMMENDED ACTIONS:

1. Check power to device
2. Verify status lights on device
3. Verify reception of packets
4. Verify network address in device and SPECTRUM
5. Cycle power on device and recheck
6. If above fails, call repair

It is useful to experiment with the alarms view by purposely causing an alarm and watching what happens. If you unplug a workstation that is connected to the network, you will see a corresponding alarm pop up in the alarms view. The new alarm will be the first one in the list, and the top portion of the screen will be dedicated to this alarm. Furthermore, if you open a network topology view you will see the workstation icon flashing red.

It is more interesting to see what happens in the alarms view if you were to unplug a router. In this event, an alarm would pop up for the router, but you would not see alarms for the devices that are unreachable by the NMP. If you opened a topology view you would see the router icon flashing red, but all the devices that are unreachable by the NMP would be gray. This means that the alarm conditions of unreachable devices are unknown.

6.1.2 The Spectrum Alarm Monitor

Next, let us experiment with the Spectrum Alarm Monitor—the ATTG component of the architecture. The function of the alarm monitor is to make alarm information available as values of parameters that can be forwarded to external applications. Once we set up the alarm monitor, we will repeat the exercise above and look at these parameters.

If you navigate to the directory SG-Tools in the Spectrum directory structure, you will see the following files (among others):

- AlarmMonitor;
- alarmrc.dat;
- SetScript;
- ClearScript;
- UpdateScript.

These files constitute the Alarm Monitor. The AlarmMonitor file is a binary executable that acts like an alarm daemon. It watches and reports all alarm activities that occur in Spectrum.

The file alarmrc.dat is the resource file for AlarmMonitor. The content of alarmrc.dat is as follows:

```
VNM NAME=my_host
TIMEOUT=120000
SOCKET NUMBER=0xBEEF
SET SCRIPT=SetScript
CLEAR SCRIPT=ClearScript
UPDATE SCRIPT=UpdateScript
```

The value of the parameter VNM NAME indicates the host on which the virtual network machine (VNM) resides. In Spectrum terminology, the

VNM is also called the SpectroSERVER. SpectroSERVER is where the network is modeled.

The values of the parameters SET SCRIPT, CLEAR SCRIPT, and UPDATE SCRIPT indicate the UNIX scripts that are executed whenever an alarm occurs, clears, or is updated.

Let us look at the content of SetScript:

```
#!/bin/sh
echo "Alarm Notification from SPECTRUM"
echo "Alarm SET:"
echo "Date:  " $1
echo "Time:  " $2
echo "Mtype:  " $3
echo "ModelName:  " $4
echo "AlarmID:  " $5
echo "Condition:  " $6
echo "CauseCode:  " $7
echo "RepairPerson:  " $8
echo "AlarmStatus:  " $9
```

This script simply echoes the parameters of each alarm as it occurs. $1 holds the date of the alarm, $2 the time, $3 the device type (or the model type), and so on.

Now we are ready to experiment with the Alarm Monitor. With the alarms view in one window, open another window and execute the following command:

```
AlarmMonitor alarmrc.dat
```

We will see a block of information appear in the window for each alarm that Spectrum knows about. Below is an example of the output:

```
Alarm Notification from SPECTRUM
Alarm SET:
Date: 08/02/94
Time: 08:05
Mtype: GnSNMPDev
ModelName: appn19d
AlarmID: 201808
Condition: RED
CauseCode: 10009
```

RepairPerson:
AlarmStatus:

Alarm Notification from SPECTRUM
Alarm SET
Date: 08/04/94
Time: 09:11
Mtype: WS_SGI
ModelName: stlucia
AlarmID: 319321
Condition: RED
CauseCode: 10009
RepairPerson:
AlarmStatus:

Alarm Notification from SPECTRUM
Alarm SET:
Date: 07/29/94
Time: 11:02
Mtype: CtDecNetApp
ModelName: DECnet Routing
AlarmID: 395
Condition: YELLOW
CauseCode: 3c000
RepairPerson:
AlarmStatus: Status Unknown

If you look at the list of outstanding alarms in the network alarms View, you will see a corresponding representation for each alarm in the UNIX window.

When all alarms have been echoed on the screen, AlarmMonitor waits for other activity on the network. Now, if we repeat the experiment of unplugging a host, we will see an additional block of information appear in the window for that alarm, and if we replug the host, we will see a block of information reporting that the alarm has been cleared.

It is important to note that the Alarm Monitor is our way of passing alarm information from Spectrum to an external application. Instead of echoing blocks of information in a window, we will pass the parameters $1, $2,..., $9 to a program that generates a trouble ticket in an external TTS. Before we do this, however, let us show two more things.

Getting Preliminary Solutions from Spectrum

First, note that the value for $7—the CauseCode—is 10009. This number represents the name of a file that contains text describing the symptom/probable cause for the alarm. To see this, navigate to the directory SG-Support/CsPCause and look at its contents. The file that contains the textual description for alarm 10009 is the file Prob00010009. The content of Prob00010009 is this:

CONTACT LOST

SYMPTOMS:

Device has stopped responding to polls

PROBABLE CAUSES:

1. Device Hardware Failure
2. Cable between this and upstream device broken
3. Power Failure
4. Incorrect Network Address
5. Device Firmware Failure

RECOMMENDED ACTIONS:

1. Check power to device
2. Verify status lights on device
3. Verify reception of packets
4. Verify network address in device and SPECTRUM
5. Cycle power on device and recheck
6. If above fails, call repair

This is the same text that would appear if you were to explode the symptom/probable cause window in Figure 6.1. The corresponding textual descriptions for other alarm types are also contained in this directory. Although the text is determined *a priori* by system designers, it is modifiable in the field. Thus, these are preliminary solutions to network alarms.

Alarm Filtering

Next, observe that we can filter out any alarms that we are not interested in. For example, suppose that we are interested only in alarms for our hubs and Cisco routers. We would add a filter that causes just the alarms on those devices to be reported, while any other alarm is discarded.

It is a simple matter to configure the filtering mechanism for different interests. We can filter with respect to alarm severity (Condition), particular network devices (ModelName), particular alarm types (CauseCode), or various combinations of these. In addition, we can enhance the mechanism in order to set alarm policies. These possibilities are left as exercises. Finally, note that this is an example of alleviating the alarm flooding problem with the ATTG component in addition to the NMP component.

Now let us revise SetScript so that it performs these two additional functions. The new SetScript retrieves the text that describes an alarm and filters out all alarms except those for hubs and Cisco routers. The revised SetScript looks like this:

```
#!/bin/sh

case "$3" in
    "HubCSIEMME" | "Rtr_CiscoMIM3T")      /* filter alarms with respect to $3 */
    pctext='cat <pathname>/Prob*"$7"' /* retrieve text, assign text to pctext */
    echo "Alarm Notification from SPECTRUM"
        .
        .
        .
    echo "CauseCode: " $7
    echo "CauseText:" $pctext      /* echo the text */
        .
        .
        .
  *) exit ;;
esac
```

Now, if we restart AlarmMonitor we will see alarms only for hubs and routers, and we will see the text that describes the alarms. For example:

Alarm Notification from SPECTRUM
Alarm SET:

   .

   .

   .

CauseCode: 1000c
  CauseText:
    BAD PORT STATUS
    A port in this device has a bad internal link status.
    PROBABLE CAUSES:

1) Loose cable connection.
RECOMMENDED ACTIONS:
1) Check the cable connections.

Alarm Notification from SPECTRUM
Alarm SET:

　·

　·

　·

CauseCode: 10009
 CauseText:
   CONTACT LOST
   SYMPTOMS:
   Device has stopped responding to polls
   PROBABLE CAUSES:
   1) Device Hardware Failure
   2) Cable between this and upstream device broken
   3) Power Failure
   4) Incorrect Network Address
   5) Device Firmware Failure
   RECOMMENDED ACTIONS:
   1) Check power to device
   2) Verify status lights on device
   3) Verify reception of packets
   4) Verify network address in device and SPECTRUM
   5) Cycle power on device and recheck
   6) If above fails, call repair

6.1.3 The Action Request System

The next stage of our tutorial is to pass the alarm parameters to a program that generates a corresponding trouble ticket in the ARS. First, we will have to devise an appropriate trouble ticket structure in the ARS. Next, we will have to write a macro that creates an instance of the ticket with the appropriate values. Finally, we will execute the macro at the command line in order to submit a trouble ticket to the TTS.

The trouble ticket is developed using the ARS Administrator Tool. First, we create a new ticket schema named the Spectrum trouble ticket, as shown in Figure 6.2.

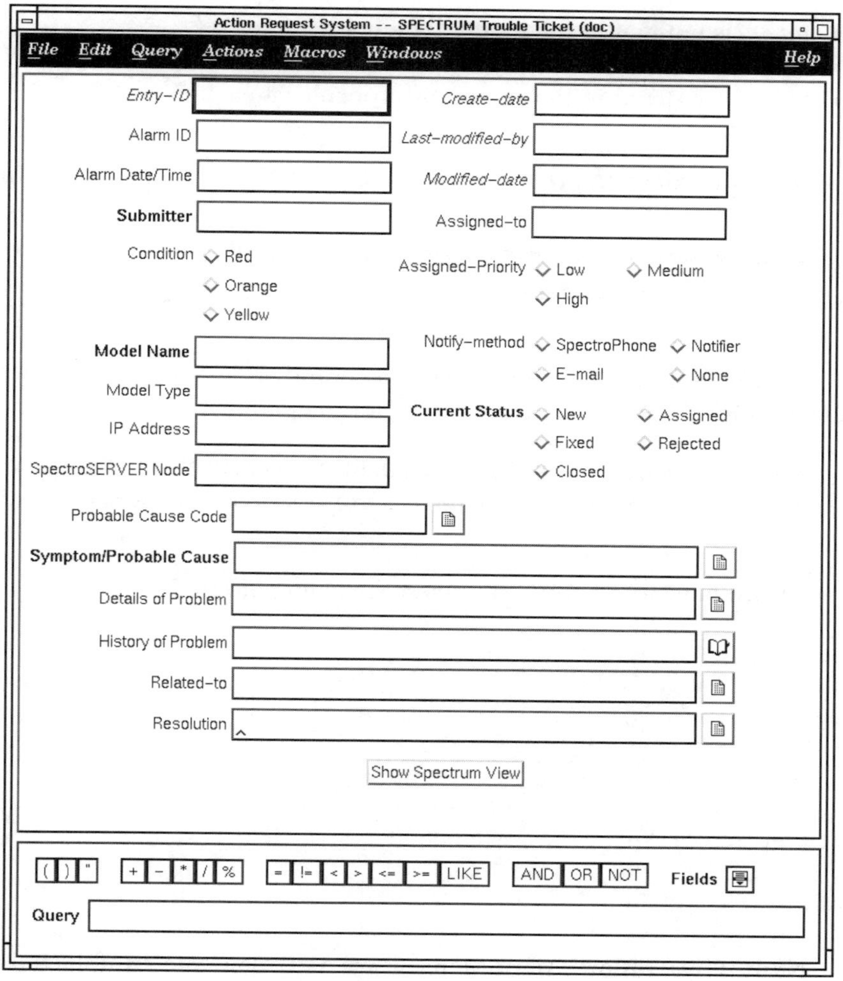

Figure 6.2 The Spectrum trouble ticket.

The macro is developed using the ARS User Tool. We create a macro named *submitTT* that takes the following variables in the ticket fields:

- *aid*   is entered in the alarm ID field;
- *cond* is entered in the condition field;
- *adt*   is entered in the alarm date/time field;
- *mn*   is entered in the model name field;

- *mt*   is entered in the model type field;
- *cc*   is entered in the probable cause code field;
- *ct*   is entered in the symptom/probable cause field.

Now, the command that generates a trouble ticket in the ARS is this:

aruser -e submitTT -p aid=$5 -p cond=$6 -p adt=$1$2 -p mn=$4 -p
mt=$3 -p cc=$7 -p ct=$pctext

We can convince ourselves that the command is effective by executing it with the following dummy values for the parameters:

aruser -e submitTT -p aid=123 -p cond=RED -p adt=1/11/99 12:00:00 -
p mn=jack -p mt=sgi_workstation -p cc=1234 -p ct=helloooooooooo

All we have to do now is to insert the command in SetScript:

```
#!/bin/sh

case "$3" in
    "HubCSIEMME"|"Rtr_CiscoMIM3T")
        pctext='cat /spectrum/SG-Support/CsPCause/Prob*"$7"'
        /ars/aruser -e submitTT -p aid=$5 -p cond=$6 -p adt=$1$2 -p
mn=$4 -p mt=$3 -p cc=$7 -p ct=$pctext ;;
    *) exit ;;
esac
```

Now if we restart AlarmMonitor we will generate trouble tickets automatically for our hubs and routers. We should try the experiment of unplugging a workstation again, but remember that we have filtered out such alarms. Assuming that we will unplug an SGI workstation, let us enter "WS_SGI" in the filter in SetScript as follows:

"HubCSIEMME"|"Rtr_CiscoMIM3T"|"WS_SGI")

If we perform the experiment now, we will get a trouble ticket for the alarm on the workstation.

6.1.4 The CBR Module

The final exercise in the tutorial is to add a CBR component to the ARS. First, we will implement a simple version of CBR and then show how we can add increasing functionality. Note that we will be following Phase I of the methodology described in Section 5.5.4: Define a list of alarm types, and implement a mechanism that retrieves similar tickets by matching with respect to alarm type.

Consider the Spectrum trouble ticket in Figure 6.2. Each instance of the ticket will describe an alarm that has been detected by Spectrum, where the information surrounding the alarm has been collected and forwarded to the ARS via AlarmMonitor.

Note the two fields probable cause code and symptom/probable cause. We have described these two fields in Section 6.1.2. The value of the probable cause code is the value of the variable $7, and the value of the symptom/probable cause is the value of $pctext. The former is an identifier for an alarm type. The latter is a textual description of the alarm symptoms and may also include a list of likely causes and suggested resolutions for the alarm (if the system designer chose to provide such information).

While the symptom/probable cause field is a broadbrush explanation of the alarm, the Resolution field will contain the actual resolution of the alarm. For example, when an instance of the trouble ticket is first created, the resolution field is empty. After the alarm is resolved, the resolution may or may not agree with the suggested recommendations in the symptom/probable cause field. As a result, it is likely that the TTS database will contain a number of tickets that agree with respect to alarm type and preliminary textual descriptions but disagree with respect to actual resolutions.

We will show how one can begin incorporating CBR in the ARS by implementing a mechanism that exploits the experiences collected in the resolution field. We will add a button to the Spectrum trouble ticket that (a) retrieves similar tickets by matching with respect to the probable cause code field, and (b) displays the resolutions in these tickets. As past resolutions are used and modified by network repair persons, the experiences are entered back into the ARS database.

Let us note that this first implementation of CBR is simply the retrieval of past resolutions based on key-term matching, where the key terms are the probable cause codes defined in Spectrum. There is no adaptation or automatic execution of the proposed resolutions.

We implement the retrieval mechanism as follows. First, using the ARS User Tool, we define a macro named *search*. In plain language, the search macro says, "Show me all tickets (except this one) that (a) match with respect to the probable cause code, and (b) whose resolution is not null." We create the macro by recording a session in which the user manually performs the action and then saves the session, giving it the name *search*.

Next, we use the ARS Administrator Tool to implement the button. The button is defined as an active link, as shown in Figure 6.3. Note that the picture in Figure 6.3 contains a conditions section and an actions section. The conditions section says that an action is to be executed whenever the button named "show similar tickets" is depressed. The actions section causes the execution of the macro named *search* with the appropriate parameters.

The button now appears in the Spectrum trouble ticket, as shown in Figure 6.4. We can verify that the button is functional by opening an existing trouble ticket and depressing the button.

If there are other tickets in the database that have the same probable cause code and also have resolutions, they will be displayed. For example, Figure 6.5 shows one of four past solutions to the "contact lost" problem, whose probable cause code is 10009. By depressing the "next" button at the bottom of the ticket, the user can view the other three resolutions to this sort of problem.

This first implementation of CBR functionality in the ARS is perhaps the simplest possible design of the CBR module. The selection and implementation of additional functions such as those described in Chapters 3 and 5 are left as exercises. In the remainder of this section, we will show how one can implement a smarter retrieval mechanism.

A way to build a smarter retrieval mechanism is to consider other fields on which to match in order to retrieve more targeted resolutions. To identify other fields that are relevant to problem solving, it is necessary to define relevance rules that will guide the search for past resolutions. See Section 3.3.2 for a discussion of relevance matching. (Note: In the CBR literature, the problem of defining relevance rules often is referred to as the indexing problem.)

For example, solutions to the problem *contact lost* (i.e., probable cause code 10009) are generally different for printers than for workstations. Similarly, solutions for contact lost with a DOS-based machine are generally different from solutions for UNIX-based machines. However,

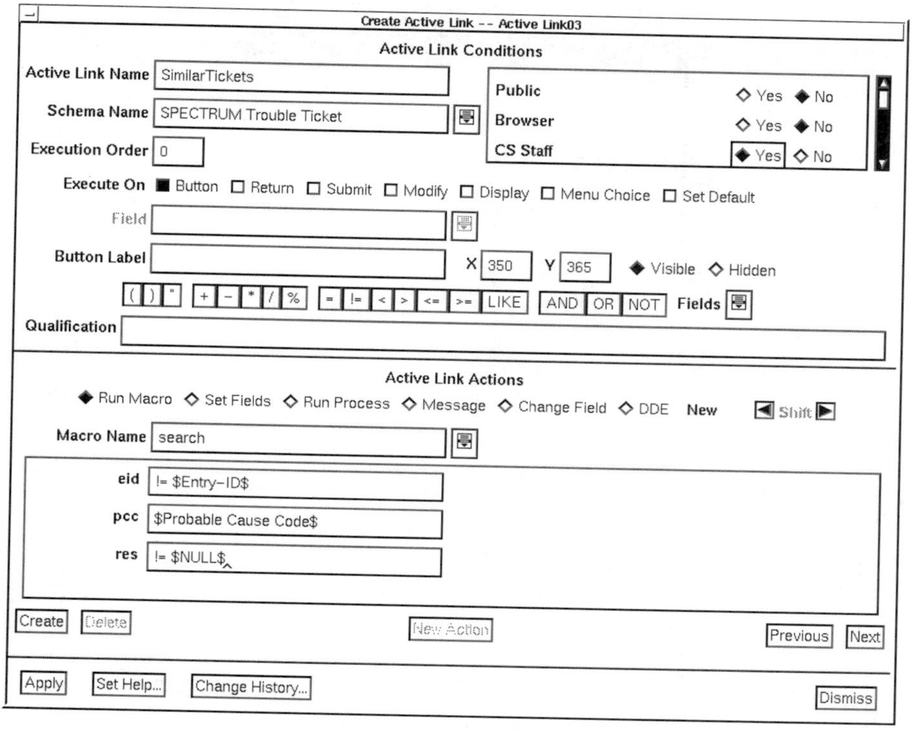

Figure 6.3   How to create a retrieve button.

given the way we have defined the retrieval mechanism above, we would retrieve past resolutions of contact lost for any type of device.

To retrieve more specific resolutions, we can reconfigure the retrieval mechanism to key off both probable cause code and model type, where model type denotes the general kind of device on which the alarm occurred. Model types are predefined in Spectrum (e.g., WS_SGI is the model type for a Silicon Graphics Workstation.) The modified retrieval mechanism says, "Show me all tickets (except this one) that (a) match with respect the probable cause code and model type, and (b) whose resolution is not null."

The next step is to follow the rest of the methodology outlined in Section 5.5.4. However, in keeping with our guiding principles (start simple and add new functionality in increments), it is wise in the beginning to implement only the retrieval mechanism described here.

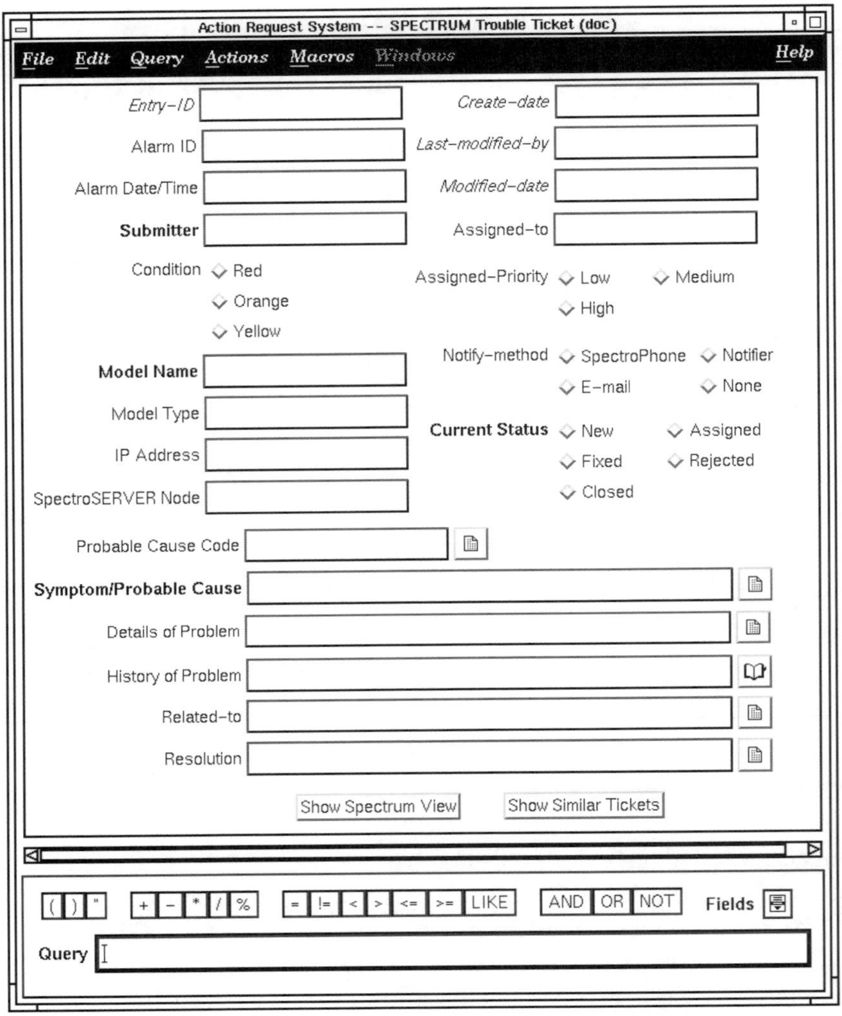

Figure 6.4 The "show similar tickets" button.

6.2 A GENERIC CBR TOOL

A limitation of the TTS/CBR approach described in Section 6.1 is that the TTS may not provide the necessary low-level functions by which to im-

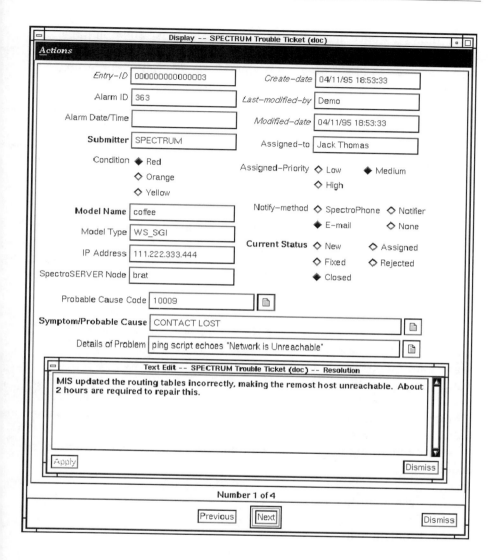

Figure 6.5 Retrieving solutions to past problems.

plement CBR functionality. Furthermore, the CBR module is more or less inextricably bound to the TTS.

In this section, we will describe a design that abstracts the features of the TTS/CBR approach. The advantage of this approach is that we separate the problem-solving experience from the TTS and, thus, generally make the experience accessible to other alarm management applications. In addition, the design affords the user the means to implement a full-featured CBR system, including increasingly complex retrieval mechanisms, adaptation, and automatic execution of proposed resolutions.

6.2.1 The Graphical User Interface

The graphical user interface of the CBR Tool consists of a primary screen that represents a case structure. The screen, as displayed in Figure 6.6, serves the following purposes:

1. It is the medium through which a case is developed and the case library is examined and expanded.
2. It is the medium through which a retrieved case for a similar problem is viewed and the solution (optionally) is adapted and executed.

The probable cause code and the symptom/probable cause are in the top section of the case structure. The probable cause code is an identifier for an alarm type. The symptom/probable cause is a textual description of the alarm type, including symptoms of the alarm, the probable cause of the alarm, and a list of recommendations for resolving the alarm. The data for these fields are predefined in Spectrum and reside in a database named CsPCause in the Spectrum directory structure.

The remaining fields in the bottom portion of the case structure are as follows:

1. Case identifier;
2. Problem key;
3. Problem text;
4. Relevant data;
5. Solution;
6. Executable solution;
7. Results;
8. Notes.

The data for these fields reside in a separate database named cbr.db. An example of an entry in cbr.db is as follows:

```
1=0000024|2=workstation_no_boot|3=tom's ibm workstation won't
boot. he was messing around with some files in
/etc/bin.|4=workstation=ibm user=tom location=build-
ing_2|5=0|6=0|7=0|8=0
```

Each instance of the case structure shown in Figure 6.6 associates an alarm type in the CsPCause database and an entry in the cbr.db database.

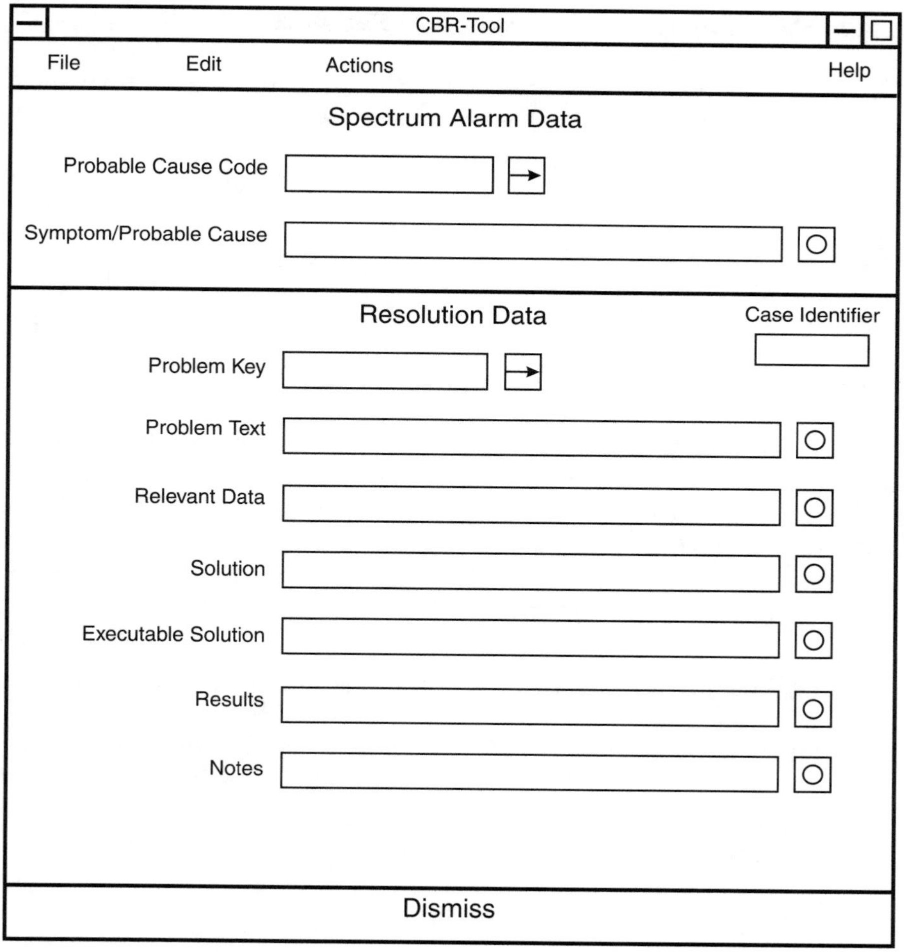

Figure 6.6 Case structure.

We should note that the CBR Tool is initiated with a case instance for each alarm type previously defined in Spectrum. However, the remaining fields 1–8 are empty. The case developer may begin building the case library by providing possible resolutions for each alarm type or may build the case library over time as the problems with specific alarms are resolved. Note that the case developers may be users of the system or designated troubleshooters.

6.2.2 Mapping Spectrum Alarms to Resolutions

Alarms and resolutions are correlated in the following way. A file named cbr.map maintains correlations between alarm types in the CsPCause database and resolutions in the cbr.db database.

An entry in cbr.map is a data structure of the following form:

<Probable Cause Code>:<Case1, Case2, ...>

For example:

```
370409:1,7,25
```

This means that the alarm type denoted by the code 370409 is associated with resolutions 1, 7, and 25. That is, the entries 1, 7, and 25 in cbr.db reflect different experiences and resolutions with regard to the 370409 alarm type in the CsPCause database.

The map and its maintenance are transparent to the user. It is maintained internally by the CBR Tool.

6.2.3 Retrieval, Adaptation, and Execution of Resolutions

Retrieval, adaptation, and execution of resolutions of alarms are achieved by invoking the CBR Tool from the command line with a select set of control options.

The command line options are these:

```
cbr    [-pcc <probable cause code>]
       [-pk <problem key>]
       [-kw <w1 w2 ...> ]
       [-kp <p1 p2 ...>
       [-exe [command]]
```

Table 6.1 provides explanations of the options:

Table 6.1
Explanations of Command Line Options

Command Line Option	Definition	Explanation
-pcc	Probable cause code	Followed by a spectrum probable cause code. Searches for cases that match with respect to the probable cause code.
-pk	Problem key	Followed by a problem key. Searches for cases that match with respect to the problem key.
-kw	Key words	Followed by a list of key words w1 w2... Searches for cases that match with respect to key words in the Problem Text field.
-kp	Key parameters	Followed by a list of parameters <p1, p2,...>. Searches for cases that match with respect to the parameters p1, p2,...in the Relevant Data field, where parameters occur on the left-hand side of the "=" sign.
-exe	Execute	Execute the command in the Executable Solution field, or else the command following -exe if one is provided. The command may take as arguments the Probable Cause Code $pcc, the problem key $pk, or the values $p1, $p2,... in the Relevant Data field. The results of the command are entered in the Results field and a record of the execution is entered in a log file.

The invocation of "cbr" with a set of control options opens the screen to 1 of M groups and 1 of N cases, where N is the number of cases in a group that satisfy the matching criteria (see Figure 6.7).

The notion of a "group" is explained below.

Groups: Suppose we want to find matches with {a b c}, including partial matches. The principle used in the CBR-Tool to find partial matches is the following. The first, most specific group of retrieved cases includes those cases that match with respect to {a b c}. The second group includes those that match with respect to {a b}, and the third group includes those that match with respect to {a}.

This grouping principle allows one to search initially for cases that are maximally similar to an input case. If good solutions are not forthcom-

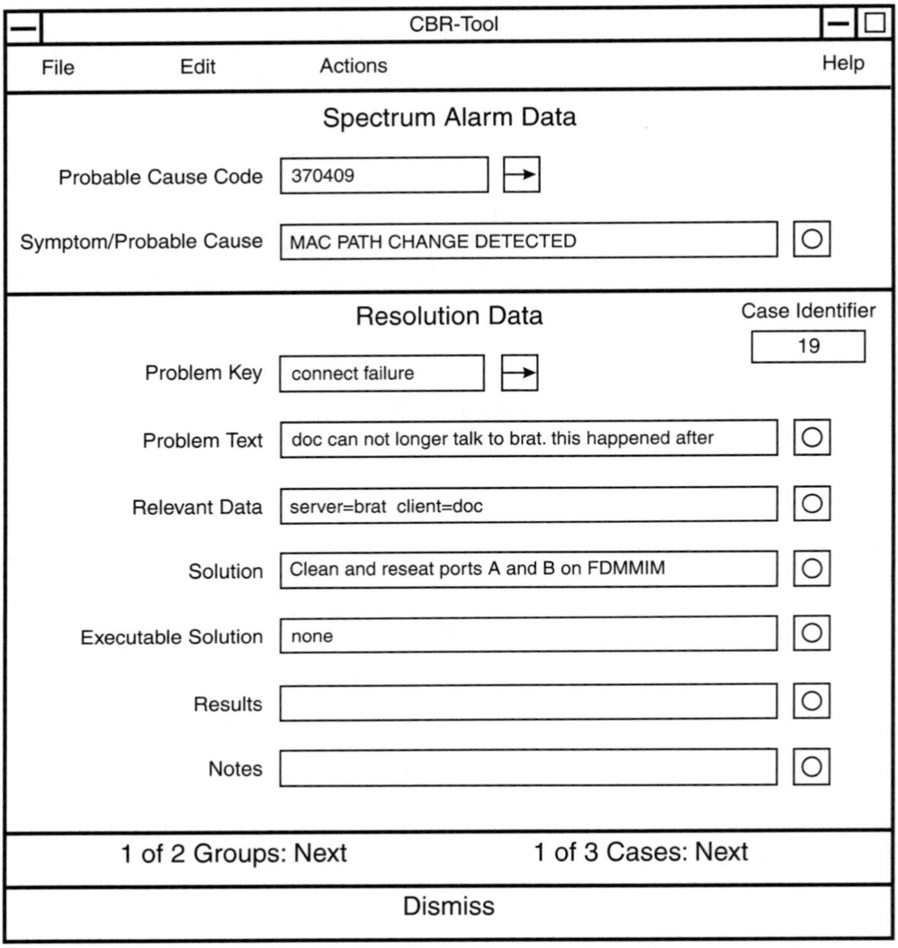

Figure 6.7 An invocation of the CBR tool.

ing, then one can examine less similar cases by clicking the "next" button (see Figure 6.7).

Also note that other grouping principles are possible. For example, one could perform matches with respect to each subset of {a b c} and, thus, retrieve a group of cases corresponding to each subset. A drawback of this principle is the problem of exponential explosion as the number of control options increase.

Note that the order in which options are listed has bearing on the groups of cases retrieved. Following are some examples of invocations of the CBR Tool with select control options:

```
cbr
```
> opens a blank case

```
cbr -pcc 370409
```
> opens to 1 of N cases with probable cause code 370409

```
cbr -kw workstation ibm
```
> opens to 1 of N cases that have the words "workstation" or "ibm" in the Problem Text field.

```
cbr -kw workstation ibm -pcc 370409
```
> opens to 1 of 2 groups and 1 of N cases that have probable cause code 370409 and have the words "workstation" or "ibm" in the Problem Text field. The second group consists of any cases that have "workstation" or "ibm" in the Problem Text field (exclusive of cases in the first group).

```
cbr -pcc 370409 -kw workstation ibm
```
> opens to 1 of 2 groups and 1 of N cases that have probable cause code 370409 and have the words "workstation" or "ibm" in the Problem Text field. The second group consists of cases that have the probable cause code 370409 (exclusive of cases in the first group).

```
cbr -pcc 370409 -kp workstation=doc -exe
```
> find the first of N cases that has (a) cause code 370409, (b) "workstation" as the left side of an "=" sign in the Relevant data field, and (c) a command in the Executable Solution field. Executes the command. Can pass the parameters $workstation, $pcc, or $pk. Log the results of the program to a log file.

```
cbr -pcc 370409 -kp workstation=doc -exe
/usr/bin/bang $pk $workstation
```
> find the first of N cases that has cause code 370409 and "workstation" as the left side of an "=" sign in the Relevant

data field, and execute the program /usr/bin/bang with the arguments $pk (i.e., whatever is in the Problem Key field) and doc (i.e., the value of $workstation). Log the results of the program to a log file.

It is important to note that the execution of a command without supervision is risky. If one wanted to find cases that have a command in the "executable solution" field but does not necessarily want to execute any one of them, one would not use the -exe option. An "execute solution" action in the actions menu of a retrieved case would be used to execute a command in the "executable solution" field of the case.

6.2.4 Using the CBR Tool

The CBR Tool is designed as a vehicle for implementing the CBR method of resolving faults in networks, where the particular CBR functionality is at the discretion of the user. The tool may be used as a retrieval mechanism by taking advantage of the pcc, pk, kw, and kp control options. In addition, CBR adaptation and execution methods are implemented by taking advantage of the exe option.

Generally, the tool can be used in the following modes:

(a) In conjunction with Spectrum alone. Following are some of the ways in which the tool is used with Spectrum:
- Example 1. Select "Show Solutions" from a menu in the alarms view in order to show past resolutions to an alarm.
- Example 2. Select "Open Case" from a menu in the alarms view in order to record a resolution to an alarm.
- Example 3. Triggered by Spectrum alarms in real time, whereupon a resolution is found and (optionally) adapted, executed, and the results entered into a log file.

(b) In conjunction with third-party alarm management systems, including trouble ticket systems, help desks, and call-tracking systems. Following are some of the ways in which the tool can be used with third-party applications.
- Example 1. Select a "Show Solutions" button or menu pick for an outstanding problem in the third-party application.

- Example 2. Triggered by the transmission of alarm information to the third-party application, whereupon a resolution is found and (optionally) adapted, executed, and reported to the user.

6.3 SUMMARY

In this chapter, we described an implementation of the fault management architecture discussed in Chapter 5. We used Spectrum and the Spectrum AlarmMonitor from Cabletron Systems as the NMP and the ATTG. We described two modes of beginning the implementation of the CBR component of the architecture. In the first mode, we used the ARS from the Remedy Corporation as the TTS and we gave a tutorial showing how one can implement a (limited) version of CBR directly into the ARS. In the second mode, we described a design for a generic CBR tool that is accessible to multiple alarm management applications.

6.4 FURTHER READING

The primary documents required to understand and implement the systems described in this chapter belong to the Spectrum suite of operation manuals, particularly:

- *The Spectrum System User's Guide*;
- *How to Integrate Applications with Spectrum*;
- *The Spectrum Alarm Monitor*;
- *The Spectrum Alarm Notifier*;
- *The Spectrum Command Line Interface*;
- *ARS Gateway User's Guide*.

Note: The Spectrum Alarm Notifier includes policy-based alarm management, alarm filters, and alarm notification in a distributed network management environment.

Additional documents required for the tutorial in Section 6.1 belong to the Action Request System suite of operation manuals:

- The ARS Administrator Tool;
- The ARS User Tool.

Epilogue

Nobody likes to spend excessive time maintaining a network, and nobody likes it when the network breaks down. One would rather have a network that maintains itself more or less autonomously.

The idea of an autonomous network has caught on in the networking community. Such networks have been referred to as intelligent networks, self-healing networks, or self-correcting networks.

The general idea of transferring a human task to some mechanical or software apparatus is the subject of this book. Specifically, the goal is to transfer and embed problem-solving expertise into existing network management software.

The model of human expertise is case-based reasoning. CBR exhitits some degree of learning, adaptability, and increasing robustness as it deals with problems in its domain. These characteristics are necessary for managing today's networks.

The book describes the tasks of network management, introduces the CBR framework of problem solving and various options for implementing CBR systems, and shows how to integrate CBR methods into existing management software. The guiding principle is to start simple and to add increasingly complex CBR methods in increments.

The introduction of CBR into network management software is new. Although there have been promising results, there is room for further advancement. The author hopes that the book has planted some good seeds, and that the approach will take on a life of its own and flourish.

List of Acronyms

API	application programming interface
ATTG	automatic trouble ticket generation
CBR	case-based reasoning
CLI	command line interface
ES	expert system
FAD	functions as designed
IAAI	Innovative Applications of Artificial Intelligence
ISO	International Standards Organization
IT	information-theoretic
MIB	management information base
NMP	network management platform
RMON	remote monitoring
SNMP	simple network management protocol
TTS	trouble ticket system
VNM	virtual network machine
WM	working memory

About the Author

Lundy Lewis is a research engineer in the Advanced Network Applications Group of Cabletron Systems, New Hampshire. As such, he designs and develops commercial products for the networking community. He has served as a team participant, project leader, and designer for several of Cabletron's network management products, including Spectrum and the ARS Gateway.

Lundy holds a B.S. in Mathematics and a B.A. in Philosophy from the University of South Carolina, an M.S. in Computer Science from Rensselaer Polytechnic Institute, and a Ph.D. in Philosophy from the University of Georgia. He is an adjunct lecturer at the University of New Hampshire, Rivier College, and New Hampshire College, where he teaches courses in artificial intelligence, neural networks, expert systems, and object-oriented design and programming. He is a member of IEEE, AAAI, and ACM.

Index

The Artech House Telecommunications Library

Vinton G. Cerf, Series Editor

For further information on these and other Artech House titles, contact:

Artech House
685 Canton Street
Norwood, MA 02062
617-769-9750
Fax: 617-769-6334
Telex: 951-659
email: artech@world.std.com

Artech House
Portland House, Stag Place
London SW1E 5XA England
+44 (0) 171-973-8077
Fax: +44 (0) 171-630-0166
Telex: 951-659
email: bookco@artech.demon.co.uk